[redacted]

Exponentially Small Splitting of Invariant Manifolds of Parabolic Points

MEMOIRS
of the
American Mathematical Society

Number 792

Exponentially Small Splitting of Invariant Manifolds of Parabolic Points

Inmaculada Baldomá
Ernest Fontich

January 2004 • Volume 167 • Number 792 (second of 5 numbers) • ISSN 0065-9266

American Mathematical Society
Providence, Rhode Island

2000 *Mathematics Subject Classification.*
Primary 37J45; Secondary 70K44, 34C37, 34E10, 37C29.

Library of Congress Cataloging-in-Publication Data

Baldomá, Inmaculada, 1971–
Exponentially small spitting of invariant manifolds of parabolic points / Inmaculada Baldomá, Ernest Fontich.
p. cm. — (Memoirs of the American Mathematical Society, ISSN 0065-9266 ; no. 792)
"Volume 167, number 792 (second of 5 numbers)."
Includes bibliographical references.
ISBN 0-8218-3445-2 (alk. paper)
1. Nonholonomic dynamical systems 2. Hamiltonian systems. 3. Lagrangian points. I. Fontich, Ernest, 1955– II. Title. III. Series.

QA3.A57 no. 792
[QA614.833]
510 s—dc22
[515′.39] 2003061939

Memoirs of the American Mathematical Society

This journal is devoted entirely to research in pure and applied mathematics.

Subscription information. The 2004 subscription begins with volume 167 and consists of six mailings, each containing one or more numbers. Subscription prices for 2004 are \$583 list, \$466 institutional member. A late charge of 10% of the subscription price will be imposed on orders received from nonmembers after January 1 of the subscription year. Subscribers outside the United States and India must pay a postage surcharge of \$31; subscribers in India must pay a postage surcharge of \$43. Expedited delivery to destinations in North America \$35; elsewhere \$130. Each number may be ordered separately; *please specify number* when ordering an individual number. For prices and titles of recently released numbers, see the New Publications sections of the *Notices of the American Mathematical Society.*

Back number information. For back issues see the *AMS Catalog of Publications.*

Subscriptions and orders should be addressed to the American Mathematical Society, P. O. Box 845904, Boston, MA 02284-5904, USA. *All orders must be accompanied by payment.* Other correspondence should be addressed to 201 Charles Street, Providence, RI 02904-2294, USA.

Memoirs of the American Mathematical Society is published bimonthly (each volume consisting usually of more than one number) by the American Mathematical Society at 201 Charles Street, Providence, RI 02904-2294, USA. Periodicals postage paid at Providence, RI. Postmaster: Send address changes to Memoirs, American Mathematical Society, 201 Charles Street, Providence, RI 02904-2294, USA.

This publication is indexed in *Science Citation Index*®, *SciSearch*®, *Research Alert*®, *CompuMath Citation Index*®, *Current Contents*®/*Physical, Chemical & Earth Sciences.*
Printed in the United States of America.

∞ The paper used in this book is acid-free and falls within the guidelines established to ensure permanence and durability.
Visit the AMS home page at http://www.ams.org/

10 9 8 7 6 5 4 3 2 1 09 08 07 06 05 04

Contents

Abstract

We consider families of one and a half degrees of freedom Hamiltonians with high frequency periodic dependence on time, which are perturbations of an autonomous system.

We suppose that the origin is a parabolic fixed point with non-diagonalizable linear part and that the unperturbed system has a homoclinic connection associated to it. We provide a set of hypotheses under which the splitting is exponentially small and is given by the Poincaré-Melnikov function.

Received by the editor April 15, 2002.

2000 *Mathematics Subject Classification.* Primary 37J45; Secondary 70K44, 34C37, 34E10, 37C29.

The authors thank the support of the Catalan Grant CIRIT 2001SGR–70.

The second author also thanks the partial support of the Spanish Grant DGICYT BFM2000–0805 and the INTAS project 00-221.

Introduction

Consider a system with an invariant object (fixed point, periodic orbit, etc) which has stable and unstable invariant manifolds associated to it and they coincide, or some branches of them coincide. If we perturb the system, generically the invariant manifolds will not coincide any more. This phenomenon is known as splitting of separatrices or splitting of invariant manifolds. One of the simplest settings where this phenomenon occurs is in differential equations in the plane having a hyperbolic saddle fixed point and a homoclinic connection associated to it. When we perturb this system with a time periodic perturbation, say

$$\dot{z} = f(z) + \varepsilon g(z, t, \varepsilon), \qquad z \in U \subset \mathbb{R}^2,$$

the fixed point becomes a hyperbolic periodic orbit with two dimensional stable and unstable invariant manifolds in $\mathbb{R}^3$. Using a first order perturbation theory the distance between the splitted manifolds measured in a plane $\{t = t_0\}$ orthogonally to the unperturbed homoclinic connection at some point p is given by

$$d(t_0, \varepsilon) = \frac{M(t_0)}{\|f(p)\|}\varepsilon + O(\varepsilon^2) \tag{1}$$

where $M(t_0)$ is the so called Poincaré-Melnikov function which is given through an integral in terms of the system and the unperturbed homoclinic orbit. For systems with slow dynamics such as

$$\dot{z} = \varepsilon f(z) + \varepsilon^2 g(z, t, \varepsilon)$$

we can scale time through $\varepsilon t = \tau$ and we obtain

$$\dot{z} = f(z) + \varepsilon g(z, \tau/\varepsilon, \varepsilon)$$

which is a perturbation of $\dot{z} = f(z)$. The formal substitution of g into the Poincaré-Melnikov function gives an ε-dependent function which is exponentially small in ε [**Fo2**]. Then, in (1) the $O(\varepsilon^2)$ term dominates over $\varepsilon M(t_0)/\|f(p)\|$ and we do not have an asymptotic expression of $d(t_0, \varepsilon)$. We only know that it is $O(\varepsilon^2)$.

One way to obtain rigorous asymptotic expressions is to introduce another parameter, that is, to consider

$$\dot{z} = f(z) + \mu g(z, t/\varepsilon, \mu).$$

Then

$$d(t_0, \mu, \varepsilon) = \frac{M(t_0, \varepsilon)}{\|f(p)\|}\mu + O(\mu^2)$$

and hence, if μ is small enough compared with $M(t_0, \varepsilon)$, which is exponentially small in ε, we have that $d \sim M(t_0, \varepsilon)\mu/\|f(p)\|$. But this only gives rigorous results in a very narrow set in the space of parameters.

Poincaré found these exponentially small effects in [**Po**]. In his study of periodic orbits in two degrees of freedom Hamiltonian systems he proposed a model, which after reduction became the following perturbed pendulum

$$\ddot{y} = 2\mu \sin y + 2\mu\varepsilon \cos y \cos t$$

(using the same notation as Poincaré). He deduced that the splitting of separatrices is exponential small in μ, provided that ε is less than an exponentially small quantity.

Arnold [**Ar1**] found the exponentially small splitting of separatrices associated to partially hyperbolic tori, studying the diffusion of action variables in near integrable systems $h_0(I) + \varepsilon h_1(\varphi, I, \varepsilon)$.

Neishtadt [**Ne**] gave upper bounds for the splitting in one and a half and two degrees of freedom Hamiltonian systems with only one parameter.

Differential equations with slow dynamics are related with near the identity diffeomorphisms by means of the Poincaré map. It turns out that being Hamiltonian is very important to get exponentially smallness. The Hamiltonian character of the equation is translated to the symplectic character of the maps.

In [**La1**] Lazutkin studied the standard map $F(x, y) = (x+y+\varepsilon \sin x, y+\varepsilon \sin x)$ and provided the following formula for the angle between the stable and unstable manifolds at a homoclinic point

$$\varphi = \frac{\pi}{\varepsilon}|\Theta_1|e^{-\pi^2/\sqrt{\varepsilon}}[1 + O(\varepsilon^b)] \tag{2}$$

with $0 < b < 1/8$. This was the first exponentially small asymptotic formula for a nontrivial problem with only one parameter. Although the proof was not complete Lazutkin introduced pioneering new analytic tools for the study of the separatrix splitting which have decisively influenced the development of this area.

Several papers deal with the computation of the constant Θ_1 [**LST**] [**Su**]. The complete proof of (2) is in [**Ge4**].

Fontich and Simó [**FS1**] [**FS2**] study the splitting of separatrices for families of diffeomorphisms in a neighborhood of the identity of class C^ω and C^r respectively. Under fairly general hypotheses exponentially small upper bounds are obtained for the distance between invariant manifolds in the analytic case with generally optimal values of the constant in the exponent.

Other works referring to maps are [**Ch**] [**DR2**] [**DR1**] [**Ge5**] [**GS**].

Many authors have studied the phenomenon of separatrix splitting with fast frequency periodic perturbations, in order to prove that in certain cases the Melnikov function yields the right asymptotics of the measure of separatrix splitting.

They consider

$$\dot{z} = f(z) + \mu\varepsilon^p g(x, t/\varepsilon, \mu), \qquad z \in U \subset \mathbb{R}^2,$$

where μ and $\varepsilon > 0$ are parameters a priori independent and such that the origin is a saddle-type fixed point. There has been a lot of discussion about the optimal value of p for which one gets exponentially small upper bounds or asymptotics. In [**Fo2**] upper bounds for the splitting are given even for negative values of p, specifically $p > -1/2$. If the model is simplified, considering equations of the form

$$\ddot{x} + f(x) = \mu\varepsilon^p g(x, t/\varepsilon, \varepsilon, \mu), \qquad x \in V \subset \mathbb{R}, \tag{3}$$

then in [**Fo1**] upper bounds are given for the splitting of separatrices for values of $p > -2$. To ask the perturbation to have order $\mu\varepsilon^p$, with p bigger than some value, depending on the method, has the advantage to provide control on the remainders in the Poincaré-Melnikov asymptotic formula. In general, if p is small, the Melnikov function does not give the right asymptotics in the case of exponentially small splitting. In [**HMS**], Holmes et al are able to give upper and lower bounds for the splitting of separatrices for quite general systems and for values of $p > 8$. The situation improves when dealing with specific systems. The most studied example is the pendulum. In [**Ge1**] and in [**DS1**], asymptotic expressions are given for the separatrix splitting of the equation

$$\ddot{x} + \sin x = \mu\varepsilon^p \sin t/\varepsilon$$

for $p > 5$ and $p > 0$ respectively. Later on, Delshams and Seara in [**DS2**], could get an asymptotic expression of the separatrix splitting for more general systems given that p is bigger than a certain quantity which depends on the perturbation and of the singularity order of the homoclinic orbit. Gelfreich in [**Ge2**] also gives an asymptotic expression for the separatrix splitting, but it is difficult to find out which p is needed in order to apply it. Finally, in [**Ge3**] Gelfreich studies in some specific examples the $p < 0$ case. The proposed method is the use of an auxiliary system whose invariant manifolds are a good approximation near the singularities of the invariant manifolds of the initial system. In [**An**] Angenent studies the splitting using variational methods. Treshev [**Tr**] studies a more general perturbation of the pendulum which includes the equation considered by Poincaré. He uses a different method based on the continuous averaging procedure developed by himself. The asymptotic formula he obtains for the area, in his example, differs from the one predicted by the Poincaré-Melnikov integral. It is worth noting that there are examples for which the asymptotic expressions are not of the form $\varepsilon^r e^{-a/\varepsilon}$, but instead involve infinitely many terms of the form $\varepsilon^{-n} e^{-a/\varepsilon}$, $n > 0$, [**SMH**].

In all these cases, one deals with Hamiltonian systems of one and a half degrees of freedom or area-preserving maps such that the origin is a hyperbolic fixed point of the non-perturbed Hamiltonian. Another situation where the separatrix splitting phenomenon appears is when one considers quasi-periodic perturbations. We refer to [**DG**], [**DGJS1**], [**DGJS2**] and [**GGM**] for such case.

Exponentially small phenomena are also found by Fiedler and Scheurle [**FS**] in one step discretizations of autonomous equations.

This memoir is devoted to study the splitting for one and a half degrees of freedom Hamiltonian systems of the form (3) such that the origin is a parabolic fixed point. Specifically we assume that the linear part of the vector field at $(0, 0)$ is

$$\begin{pmatrix} 0 & 1 \\ 0 & 0 \end{pmatrix}.$$

We consider the case of fast frequency perturbation. The paper [**CFN**] deals with the case of constant frequency. The first point is to put sufficient conditions such that the perturbed system also has invariant manifolds.

We have followed basically the structure of [**DS2**]. However, due to the fact that many of their arguments strongly rely on the hyperbolic character of the fixed point, we have had to introduce new techniques to deal with the parabolic case. To this end we have also used tools introduced by Lazutkin [**La2**] [**La1**]. It is worth remarking that most of our arguments are can be adapted for the hyperbolic case.

The memoir is organized as follows. In the first chapter we introduce the notation, the hypotheses and the main theorem.

In the second chapter, we study some analytical properties of the homoclinic orbit of the unperturbed system. In particular we get the asymptotic behavior of its parameterization. We prove that, as was to be expected, this behavior is algebraic, that is, there exists $T > 0$ such that if $t \in \mathbb{C}$, $\operatorname{Re} t \geq T$, the stable manifold behaves like $1/t^p$ with p a certain positive number, and, analogously, the unstable manifold has the form $1/(-t)^p$ for $t \in \mathbb{C}$, $\operatorname{Re} t < -T$.

In the third chapter, we establish, under the stated conditions, the existence of stable and unstable invariant manifolds for the perturbed system. Moreover we find useful parameterizations $\gamma^*(t,s)$ $(* = \mathrm{s},\mathrm{u})$, of the local invariant manifolds of the perturbed Hamiltonian system. These parameterizations satisfy that γ^* is a solution with respect to the variable $t \in \mathbb{R}$, it is analytic with respect to s and

$$\gamma^*(t + 2\pi\varepsilon, s) = \gamma^*(t, s + 2\pi\varepsilon).$$

In this way we endow the variable s with a dynamic character since, if P^{t_0} is the Poincaré map from $t = t_0$ to $t = t_0 + 2\pi\varepsilon$, $\gamma^{\mathrm{s}}(t_0, s)$ represents the stable manifold of P^{t_0} and the dynamics of P^{t_0} on it is simply

$$P^{t_0}(\gamma^{\mathrm{s}}(t_0, s)) = \gamma^{\mathrm{s}}(t_0, s + 2\pi\varepsilon).$$

Moreover γ^* is of the form

$$\gamma^*(t,s) = \gamma_0(t+s) + \mu\varepsilon^{p+1}\sigma^*(t,s),$$

where γ_0 is the homoclinic orbit of the unperturbed system.

In the fourth chapter, we built the flow box coordinates, i.e., coordinates in which the flow straightens. These coordinates are defined in a neighborhood of the stable manifold not containing the origin, but close to it and independent of the parameters. We built them following several steps. We parameterize the solutions of the perturbed system near a piece of the stable manifolds by two parameters. One of these is time, and the other is a complex parameter s such that the solutions are analytic with respect to s and the dynamics of the Poincaré mapping is simply $s \mapsto s + 2\pi\varepsilon$. We can write them in the form $w(t+s, t/\varepsilon)$. We prove afterwards, thanks to this good parameterization, that the solutions intersect a (real) section transverse to the flow for some value (t_0, s_0), thus we are able to straighten the flow in a neighborhood of the stable manifold. Finally we slightly modify the variables to make them canonical.

In the fifth chapter, we present a result of Delshams-Seara [**DS2**] which asserts that if p is bigger than some value, which depends on the perturbation and the unperturbed homoclinic orbit, we can extend the parameterization of the unstable manifold until it reaches the domain where the flow box variables are defined.

Finally, in the last chapter we introduce the splitting function. From it and its properties we derive the asymptotic formulas for the area of the lobes generated by the invariant manifolds between two homoclinic points and the angle between the invariant manifolds at a homoclinic point. They are exponentially small in ε. The main difference with [**DS2**] is that here we consider homoclinic orbits with algebraic branch type singularities and consequently some computations are somewhat more involved.

1. Notation and main results

In this chapter we present the main problem we consider, the hypotheses we assume, and the rigorous statement of the main results. For that we have to begin by introducing some notation.

At the end we present an example where the above mentioned results apply.

1.1. Notation and hypotheses

We study the splitting of separatrices in the case which we call the parabolic case. Next we describe the settings of this case and the hypotheses we will need.

We consider Hamiltonian systems of one and a half degrees of freedom with Hamiltonian

$$H(x, y, t/\varepsilon, \mu, \varepsilon) = h_0(x, y) + \mu\varepsilon^p h_1(x, y, t/\varepsilon, \mu, \varepsilon)$$

where

$$h_0(x, y) = \frac{y^2}{2} + V(x),$$

$V(x)$ is an analytic function of order n, that is

$$V(x) = a_n x^n + \cdots$$

with $n \geq 3$. With these assumptions, for the unperturbed system (i.e. the system when $\mu = 0$) the origin is a parabolic fixed point and the derivative of the Hamiltonian vector field at $(0, 0)$ is

$$\begin{pmatrix} 0 & 1 \\ 0 & 0 \end{pmatrix}.$$

The differential equations associated to the Hamiltonian are

$$(1.1) \qquad \begin{aligned} \dot{x} &= y + \mu\varepsilon^p \partial_y h_1(x, y, t/\varepsilon, \mu, \varepsilon) \\ \dot{y} &= -V'(x) - \mu\varepsilon^p \partial_x h_1(x, y, t/\varepsilon, \mu, \varepsilon). \end{aligned}$$

We will assume the following hypotheses related to the unperturbed system. Note that the unperturbed system is autonomous and independent on ε.

1.1.1. Hypothesis for the unperturbed system.

HP1 We assume that, $h_0(x, y) = y^2/2 + V(x)$ is analytic and $V(x) = a_n x^n + \cdots$ with $a_n < 0$ and $n \geq 3$. Moreover we assume that h_0 has a homoclinic orbit, associated to the equilibrium point $(0, 0)$

We denote the time parameterization of the homoclinic orbit by

$$\gamma_0(u) = (\alpha_0(u), \beta_0(u))$$

with some chosen (fixed) initial condition $\gamma_0(0) = (x_0, y_0)$ on the homoclinic orbit.

We assume that $\gamma_0(u)$ is analytic in a complex strip $|\operatorname{Im} u| < a$ with branching points at $u = \pm ia$, i.e., there exists $\rho > 0$ such that for $u \in \mathbb{C}$

satisfying $|u-ia|<\rho$, $\arg(u-ia)\in(-3\pi/2,\pi/2)$, $\gamma_0(u)$ can be expressed as

$$\alpha_0(u)=\frac{d_-}{(u-ia)^{c/q}}(1+O(u-ia)^{1/q}),\ \beta_0(u)=\frac{e_-}{(u-ia)^{1+c/q}}(1+O(u-ia)^{1/q}).$$

and for $u\in\mathbb{C}$ such that $|u+ia|<\rho$, $\arg(u+ia)\in(-\pi/2,3\pi/2)$, $\gamma_0(u)$ can be expressed as

$$\alpha_0(u)=\frac{d_+}{(u+ia)^{c/q}}(1+O(u+ia)^{1/q}),\ \beta_0(u)=\frac{e_+}{(u+ia)^{1+c/q}}(1+O(u+ia)^{1/q}),$$

where $c,q\in\mathbb{Z}$, $q\neq 0$. Moreover on $u=\pm ia$ there are no other singularities of γ_0. We define

$$r=1+\frac{c}{q}>1.$$

Of course, poles are included in this definition of branching points.

REMARK 1.1. *According to Proposition 2.3 (Chapter 2) always exists $a>0$ such that $\gamma_0(u)$ is analytic on the strip $\{u\in\mathbb{C}:|\operatorname{Im}u|<a\}$.*

REMARK 1.2. *According to Proposition 2.4 (Chapter 2), if $V(x)=a_nx^n+\cdots+a_mx^m$ is a polynomial and we assume that $\alpha_0(u)$ has a singularity at $u=u^*\in\mathbb{C}$, then for u in a neighborhood of u^* we have that*

$$\begin{aligned}\alpha_0(u)&=\frac{C}{(u-u^*)^{2/(m-2)}}(1+O(u-u^*)^{2/(m-2)})\\ \beta_0(u)&=-\frac{C'}{(u-u^*)^{m/(m-2)}}(1+O(u-u^*)^{2/(m-2)}).\end{aligned}$$

As a consequence, the exponents of $u-u^*$ in the expressions of α_0 and β_0 are rational numbers.

1.1.2. Hypotheses over the perturbation.

HP2 The function $h_1(x,y,\theta,\mu,\varepsilon)$ is defined for $(x,y)\subset U\subset\mathbb{C}^2$, $\theta\in\mathbb{R}$, $\mu\in\mathbb{D}(0,\mu_0)$, $\varepsilon\in(0,\varepsilon_0)$, it is C^0 and 2π-periodic in θ, has zero mean:

$$\int_0^{2\pi}h_1(x,y,\theta,\mu,\varepsilon)\,d\theta=0$$

and it is real analytic with respect to (x,y,μ).

HP3 The function $h_1(x,y,\theta,\mu,\varepsilon)$ is a polynomial of degree κ and order k (i.e. the lowest degree of the monomials in h_1) in the (x,y) variables. That is

$$h_1(x,y,\theta,\mu,\varepsilon)=\sum_{i+j=k}^{\kappa}a_{i,j}(\theta,\mu,\varepsilon)x^iy^j.$$

HP4 The order k of the perturbation satisfies

$$2k-2\geq n.$$

REMARK 1.3. *We observe that **HP4** implies that the origin also is a parabolic fixed point of the perturbed system and the derivative of the vector field evaluated at this point is the same as the one of the unperturbed system.*

Consider the terms $a_{i,j}(\theta,\mu,\varepsilon)x^iy^j$ of h_1 evaluated on γ_0. We define ℓ to be the greatest order of the branching points $\pm ia$ corresponding to $a_{i,j}(\theta,\mu,\varepsilon)\alpha_0^i(u)\beta_0^j(u)$. That is:

$$\ell = \max\{i(r-1)+jr : a_{i,j}(\theta,\mu,\varepsilon)\neq 0\}. \tag{1.2}$$

Also we define

$$\nu = p-\ell.$$

HP5 The constant ν is greater or equal than 0.

REMARK 1.4. *Hypothesis **HP5** controls the growth of the perturbation term*

$$\mu\varepsilon^p h_1(x,y,t/\varepsilon,\mu,\varepsilon)$$

*evaluated at the homoclinic orbit, near the singularities. In fact, if hypothesis **HP5** is assumed:*

$$\mu\varepsilon^p\|h_1(\gamma_0(u),t/\varepsilon,\mu,\varepsilon)\|_\infty = O(\mu),$$

for $|\operatorname{Im} u|\leq a-\varepsilon$.

REMARK 1.5. *According to Hypotheses **HP1-HP5**, if $p<1$, then $\partial_y h_1 = 0$. Indeed, if $\ell\geq 1$, then by hypothesis **HP5**, $p\geq 1$. Therefore, we consider the case $\ell<1$. By definition of ℓ and using that $r\geq 1$, we have that for any pair of positive integers, i, j such that $a_{i,j}(\theta,\mu,\varepsilon)\neq 0$,*

$$1>\ell\geq i(r-1)+jr\geq jr\geq j.$$

Therefore, $j=0$ and this implies that h_1 has no terms depending on the variable y. Therefore $\partial_y h_1 = 0$.

1.2. Main results

It is a well known fact that Poincaré maps associated to periodic Hamiltonian perturbations of one degree of freedom Hamiltonian systems having a homoclinic connection, have either primary homoclinic points or a homoclinic connection, the latter possibility being non-generic. These points are related to the zeros of the Melnikov function $M(s,\varepsilon)$ defined by

$$M(s,\varepsilon) = \int_{-\infty}^{\infty}\{h_0,h_1\}(\gamma_0(t+s),t/\varepsilon)\,dt.$$

Let P^{t_0} be the Poincaré map from t_0 to $t_0+2\pi\varepsilon$. We denote by A the area of the lobe generated by the stable and the unstable manifolds between two homoclinic points and by ϑ the angle between the stable and unstable invariant manifolds at a homoclinic point. We observe that, since the Poincaré map is area preserving, the area A will not depend on the homoclinic points.

The main results are:

THEOREM 1.1. *Under hypotheses **HP1-HP5**, for $\varepsilon\to 0^+$, $\mu\to 0$, and for any $t_0\in\mathbb{R}$, the following formulae hold:*

$$\begin{aligned}
A &= \mu\varepsilon^p\int_{s_0}^{\bar{s}_0} M(v,\varepsilon)\,dv + O(\mu^2\varepsilon^{2\nu+r},\mu^2\varepsilon^{\nu+p+1},\mu\varepsilon^{p+2})e^{-a/\varepsilon},\\
\sin\vartheta &= \mu\varepsilon^p\frac{M'(s_0,\varepsilon)}{\|\dot\gamma_0(t_0+s_0)\|^2} + O(\mu^2\varepsilon^{2\nu+r-2},\mu^2\varepsilon^{\nu+p-1},\mu\varepsilon^p)e^{-a/\varepsilon},
\end{aligned}$$

where $s_0 < \bar{s}_0$ are the two zeros of the Melnikov function (associated to two consecutive homoclinic points), closest to zero, which depend on t_0. Furthermore $t_0 + s_0(t_0)$ is $2\pi\varepsilon$ periodic.

REMARK 1.6. *Since $t_0 + s_0(t_0)$ is $2\pi\varepsilon$-periodic, the expression for the angle ϑ is $2\pi\varepsilon$-periodic in t_0.*

We define the function

$$J(x, y, \theta) = \{h_0, h_1\}(x, y, \theta).$$

By hypothesis **HP2** on h_1, J is 2π-periodic in θ and has zero average with respect to θ. Then we can consider its Fourier expansion

$$J(x, y, \theta) \sim \sum_{k \neq 0} J_k(x, y) e^{ik\theta}.$$

Moreover, for all $k \in \mathbb{Z}$, $J_k(\gamma_0(u))$ has a branching point of order at most $\ell + 1$ at $u = \pm ia$. Therefore, near the singularity $u = ia$, $J_k(\gamma_0(u))$ has the form

$$J_k(\gamma_0(u)) = \frac{1}{(u - ia)^{\ell+1}} \left(J_{k,0}^- + \sum_{m \geq 1} J_{k,m}^- (u - ia)^{m/q} \right)$$

and, near the singularity $u = -ia$, $J_k(\gamma_0(u))$ has the form

$$J_k(\gamma_0(u)) = \frac{1}{(u + ia)^{\ell+1}} \left(J_{k,0}^+ + \sum_{m \geq 1} J_{k,m}^+ (u + ia)^{m/q} \right).$$

We note that $J_{k,0}^+ = \overline{J_{-k,0}^-}$.

We further consider the following hypothesis:

HP6 The Fourier coefficients $J_{\pm 1}$ evaluated on $\gamma_0(u)$, that is $J_{\pm 1}(\gamma_0(u))$, have singularities of order exactly $\ell + 1$ at the points $u = \pm ai$.

REMARK 1.7. *Hypothesis **HP6** is generic because it is equivalent to suppose that the coefficients $J_{\pm 1,0}^{\mp}$ of the Laurent expansion of $J_{\pm 1}(\gamma_0(u))$ are different from zero.*

We can obtain an asymptotic expression of the Melnikov function and consequently of the area of the lobe and of the angle.

COROLLARY 1.1. *If **HP1-HP6** hold, then for $\varepsilon \to 0^+$, $\mu \to 0$ and for any $t_0 \in \mathbb{R}$,*

$$\begin{aligned}
M(s, \varepsilon) &= \varepsilon^{-\ell} \frac{4\pi}{\Gamma(\ell+1)} |J_{1,0}^-| \operatorname{Re}(e^{i(\theta - (\ell+1)\pi/2)} e^{-is/\varepsilon}) e^{-a/\varepsilon} + O(\varepsilon^{-\ell+1} e^{-a/\varepsilon}), \\
A &= \mu \varepsilon^{\nu+1} \frac{8\pi}{\Gamma(\ell+1)} |J_{1,0}^-| e^{-a/\varepsilon} + O(\mu^2 \varepsilon^{2\nu+r}, \mu^2 \varepsilon^{\nu+p+1}, \mu \varepsilon^{\nu+2}) e^{-a/\varepsilon}, \\
|\sin \vartheta| &= \mu \varepsilon^{\nu-1} \frac{4\pi}{\Gamma(\ell+1)} |J_{1,0}^-| \frac{1}{\|\dot{\gamma}_0(t_0 + s_0)\|^2} e^{-a/\varepsilon} \\
&\quad + O(\mu^2 \varepsilon^{2\nu+r-2}, \mu^2 \varepsilon^{\nu+p-1}, \mu \varepsilon^{\nu}) e^{-a/\varepsilon},
\end{aligned}$$

where $J_{1,0}^- = |J_{1,0}^-| e^{i\theta}$ and Γ is the Gamma function.

1.3. Example

An example of an unperturbed Hamiltonian system satisfying **HP1** is given by

$$h_0(x,y) = \frac{y^2}{2} + V(x) \tag{1.3}$$

where $V(x)$ is a polynomial of the form

$$V(x) = -x^n + x^{2(n-1)}, \qquad n \geq 3.$$

Indeed, the Hamiltonian system has a homoclinic orbit contained in $H(x,y) = 0$. Let $\gamma_0(t) = (\alpha_0(t), \beta_0(t))$ be the parameterization of the homoclinic orbit such that $\gamma_0(0) = (1,0)$.

We can give an explicit expression it:

$$\alpha_0(t) = \left(\frac{2}{2+(n-2)^2t^2}\right)^{1/(n-2)}, \qquad \beta_0(t) = -(n-2)t(\alpha_0(t))^{n-1}.$$

Therefore, the homoclinic orbit has singularities at the points $\pm ia$ with $a = \sqrt{2}/(n-2)$ which are branching points (if $n = 3$, are poles). It is not difficult to see that, near the singularities $\pm ia$, the first component of γ_0 reads as

$$\frac{C_\pm}{(t \pm ia)^{1/(n-2)}}\left(1 + O(t \pm ia)^{1/(n-2)}\right)$$

with $C_- = (a/2)^{1/(n-2)} e^{-i\pi/2(n-2)}$ and $C_+ = \overline{C_-}$.

We consider a family of perturbations given by

$$\mu\varepsilon^p h_1(x,y,t/\varepsilon) = \mu\varepsilon^p x^k \cos(t/\varepsilon), \qquad p \geq k/(n-2).$$

In this case, $\ell = k/(n-2)$. Of course we assume that k satisfies the hypothesis **HP4**, that is, $2k-2 \geq n$. Hence in this case $\ell \geq (n+2)/(2n-4)$.

Then, by Corollary 1.1 the area of the lobe generated between two consecutive homoclinic points satisfies the asymptotic expression

$$A \sim \mu\varepsilon^{\nu+1}\frac{4\pi}{\Gamma\left(\frac{k}{n-2}\right)}|C_-|^k e^{-a/\varepsilon}$$

where $\nu = p - k/(n-2)$.

2. Analytic properties of the homoclinic orbit of the unperturbed system

2.1. Introduction and main results

The purpose of this chapter is to obtain the asymptotic behaviour of the homoclinic orbits to parabolic points of Hamiltonians systems of the form

$$H(x,y) = y^2/2 + V(x)$$

with $V(x)$ being an analytic function, for complex values of time in a certain domain.

We assume that the origin is a fixed point of the corresponding Hamiltonian equation

$$\begin{aligned} \dot{x} &= y \\ \dot{y} &= -V'(x). \end{aligned} \tag{2.1}$$

It is not restrictive to assume that $V(0) = 0$. We suppose that V is of the form

$$V(x) = a_n x^n + \dots$$

with $n \geq 3$ and $a_n \neq 0$.

In such case the origin is a parabolic point, that is, the linear part of the equation at $(0,0)$ has a double zero eigenvalue. Assuming that the origin has an invariant curve passing through the origin, the solution on this curve has to lie on the energy level $H(x,y) = 0$. Then

$$\dot{x} = y = \pm\sqrt{-2V(x)}.$$

Hence we will have that $\dot{x} = ax^k + \dots$ or $\dot{x} = ax^{k+1/2} + \dots$ according to the cases $n = 2k$ or $n = 2k+1$, $k \in \mathbb{N}$.

For the sake of generality we consider the case $k \in \mathbb{R}$. We define the set

$$U = \mathbb{D}(0,r) \setminus \{z \in \mathbb{C} : \operatorname{Im} z = 0, \operatorname{Re} z \leq 0\} \subset \mathbb{C}.$$

The main result of this chapter is the following proposition from which we derive the asymptotic representation of $x(t)$, and then $y(t)$ follows from $y(t) = \dot{x}(t)$.

PROPOSITION 2.1. *Let f be an analytic function on U. Suppose that*

$$f(x) = ax^k + g(x),$$

with $|g(x)| \leq B|x|^\ell$, k, $\ell \in \mathbb{R}$, $1 < k < \ell$ and $a < 0$. Consider the equation

$$\dot{x} = f(x).$$

Then, there is an analytic solution $\varphi(t)$ defined on

$$\Omega(T,\alpha) = \{t \in \mathbb{C} : |t| > T, |\arg t| < \alpha\}$$

with $\alpha < \min\{\pi, \frac{\pi}{p}\}$ and T big enough, such that

$$\varphi(t) = ct^{-p} + O(t^{-\nu})$$

with $p = 1/(k-1)$, $p < \nu < \min\{q, p+1\}$, $q = p(1+\ell-k)$ and $c = (-p/a)^p$.

REMARK 2.1. *Since the equation is one-dimensional, every solution which goes to zero as* $\mathrm{Re}\, t$ *goes to* $+\infty$, *with the real part of the initial condition positive, is of the form* $\varphi(t+\tau)$.

Note that if we assume that the leading term of φ is ct^{-p} a formal computation already shows that

$$p = \frac{1}{k-1}, \qquad c = \left(\frac{-p}{a}\right)^p = \left(\frac{1}{(1-k)a}\right)^{1/(k-1)}.$$

The proof of Proposition 2.1 is given in the next section.

If we restrict us to consider functions f of the form $f(x) = x^{n/2}g(x)$, which are the ones we will deal with in the next chapters, we obtain a bigger domain for the solution.

PROPOSITION 2.2. *Let* $f(x) = x^{n/2}g(x)$, $n \in \mathbb{N}$, $g(0) < 0$ *and* g *analytic in* $\mathbb{D}(0,r)$. *Then equation* $\dot{x} = f(x)$ *has an analytic solution* φ, $\varphi(t) = ct^{-p} + O(t^{-\nu})$, *defined on* $\Omega(T,\pi)$.

The proof is almost the same as the one of Proposition 2.1. We only have to take into account that $x^{n/2}$ is continued analytically to its Riemman surface with $|x^{n/2}| \le |x|^{n/2}$.

REMARK 2.2. *According to Proposition 2.2, if system (2.1) has a homoclinic orbit,* σ, *it has not periods. This is because* $\lim_{|t|\to+\infty} \sigma(t) = 0$. *This is in contrast with the hyperbolic case where always there exists an imaginary period. (See* [**Fo2**]*)*

From the previous results we easily obtain that a homoclinic orbit to a parabolic point has singularities and they do not accumulate to the real axis:

PROPOSITION 2.3. *Let* $H(x,y) = y^2/2 + V(x)$ *be a Hamiltonian with a parabolic equilibrium point and* $\sigma(t)$ *a homoclinic orbit associated to it. Then*

1) σ *has singularities,*
2) *there exists* $\eta > 0$ *such that* σ *is analytic in the strip* $\{t \in \mathbb{C} : |\operatorname{Im} t| < \eta\}$.

PROOF. We may assume that the equilibrium point is the origin. Therefore

$$\lim_{t\to\pm\infty} \sigma(t) = 0.$$

Applying Proposition 2.2 to the equation reduced to the stable invariant manifold when $t \to +\infty$ and to the unstable one when $t \to -\infty$ (changing t by $-t$) and taking into account Remark 2.1, we obtain that there exist T, t_1 and t_2 such that

$$\begin{aligned} \sigma(t) &= \frac{c_1}{(t-t_1)^p} + O\left(\frac{1}{(t-t_1)^\nu}\right), \qquad |t| \ge T,\ |\arg t| < \pi, \\ \sigma(t) &= \frac{c_2}{(t-t_2)^p} + O\left(\frac{1}{(t-t_2)^\nu}\right), \qquad |t| \ge T,\ |\pi - \arg t| < \pi. \end{aligned}$$

If we suppose that σ is an entire function, the previous expressions imply that σ is bounded outside the disc of radius T. Then, by Liouville's theorem, it must be constant. This contradiction implies that σ has at least one singularity which has to be in $\mathbb{D}(0,T)$.

To prove the existence of $\eta > 0$ we just have to note that σ is analytic on $\mathrm{Re}\, t > T$ and on $\mathrm{Re}\, t < -T$. Since σ is analytic on $\mathbb{R}$, and therefore on $[-T,T]$, there exists $\eta > 0$ such that σ is analytic on $\{t \in \mathbb{C} : |\mathrm{Re}\, t| \le T,\ |\operatorname{Im} t| < \eta\}$. $\square$

We write $\sigma = (\alpha_0, \beta_0)$.

PROPOSITION 2.4. *Let $V(x) = a_n x^n + \cdots + a_m x^m$ be a polynomial and $u = u^*$ a singularity of σ. Then*

$$\alpha_0(u) = \frac{C}{(u-u^*)^{2/(m-2)}}(1 + O(u-u^*)^{2/(m-2)}) \tag{2.2}$$

$$\beta_0(u) = -\frac{C'}{(u-u^*)^{m/(m-2)}}(1 + O(u-u^*)^{2/(m-2)}). \tag{2.3}$$

PROOF. It is clear that α_0 is a solution of the equation $\dot{x} = \pm\sqrt{-2V(x)}$. Performing the change $w = 1/x$ we obtain the equation

$$\frac{du}{dw} = \frac{\mp w^{m/2-2}}{\sqrt{-2(a_m + wa_{m-1} + \ldots + w^{m-n}a_n)}}$$

which, in a neighborhood of $w = 0$, can be written as

$$\frac{du}{dw} = w^{m/2-2}(c_0 + O(w)).$$

Integrating this relation we obtain

$$u - u^* = w^{(m-2)/2}(c_1 + O(w)).$$

Inverting the last equation and going back to the variable x we obtain the claimed expressions (2.2) and (2.3). □

In the following proposition we will give a qualitative study of the domain $\alpha_0(\{u \in \mathbb{C} : \mathrm{Re}\, u \geq \kappa\})$.

PROPOSITION 2.5. *Let $V(x) = a_n x^n + \cdots$ be an analytic function. Then, there exists κ_0 such that, if $\kappa \geq \kappa_0$, α_0 applies the semiplane $\{u \in \mathbb{C} : \mathrm{Re}\, u \geq \kappa\}$ bijectively to a domain $D^*(\kappa)$ of $\mathbb{C}$. The image by α_0 of the line $\{\mathrm{Re}\, u = \kappa\}$ together with $\{0\}$ is a single closed curve which is the boundary of $D^*(\kappa)$. It is a smooth curve except at zero. The tangent lines near zero have limit slopes $\pm\pi/(n-2)$.*

In Figures 1 and 2 we display the domain $D^*(\kappa)$, with κ big, and $n = 2$ and $n = 7$ respectively

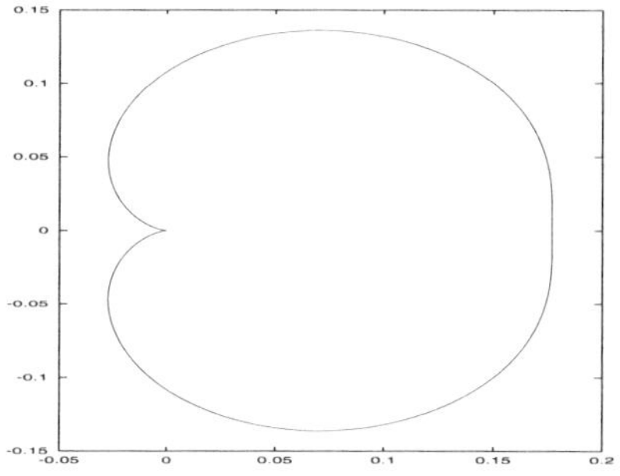

Figure 1

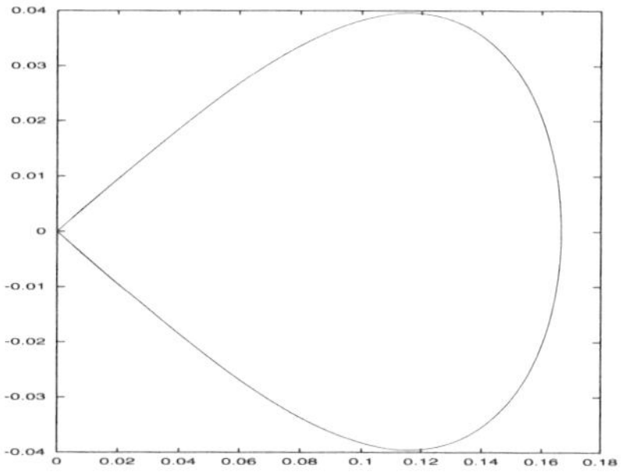

Figure 2

PROOF. We assume that there exists $u_1, u_2 \in \{\mathrm{Re}\, u \geq \kappa\}$ such that $\alpha_0(u_1) = \alpha_0(u_2)$. Then the points $(\alpha_0(u_j), \beta_0(u_j))$, $j = 1, 2$ stay in the zero energy level, and hence $\beta_0(u_1) = \pm\beta_0(u_2)$. In the case that the sign is $+$, we should have that $\alpha_0(t + u_2 - u_1) = \alpha_0(t)$ for all t in the domains of α_0, that is $u_2 - u_1$ is a period of α_0, which is a contradiction with Remark 2.2

In the case that the sign is $-$, we use that, due to the symmetry of the equation, $\alpha_0(t) = \alpha_0(u_2 + u_1 - t)$ and $\beta_0(t) = -\beta_0(u_2 + u_1 - t)$. Then $\beta_0((u_1 + u_2)/2) = 0$ and therefore, $V(\alpha_0((u_1 + u_2)/2)) = 0$. Since 0 is an isolated zero for V, if κ is big enough, $\alpha_0(\{\operatorname{Re} u \geq \kappa\})$ does not contain any zero of V, which is a contradiction.

Let

$$g(t) = \alpha_0(it + \kappa) = \frac{c}{(it + \kappa)^p} + O(|it + \kappa|^{-\nu}),$$

with $p = 2/(n-2)$. The claim on the tangent lines comes from

$$\lim_{t \to \pm\infty} \arg g(t) = \mp p \frac{\pi}{2}.$$

□

2.2. Proof of Proposition 2.1

Let $U = \mathbb{D}(0, r) \setminus \{z \in \mathbb{C} : \operatorname{Im} z = 0,\ \operatorname{Re} z \leq 0\} \subset \mathbb{C}$ and $k, \ell \in \mathbb{R}$, $1 < k < \ell$. We consider

$$f(x) = f_0(x) + g(x)$$

with $f_0(x) = ax^k$, $a < 0$, and $g : U \to \mathbb{C}$ analytic such that $|g(x)| \leq B|x|^\ell$.

A solution of

$$\dot{x} = ax^k$$

is $x(t) = ct^{-p}$ with $c = (-p/a)^p$ and $p = 1/(k-1)$. We look for a solution of $\dot{x} = f(x)$ of the form

$$\varphi(t) = \varphi_0(t) + \psi(t), \qquad \text{with} \qquad \varphi_0(t) = ct^{-p}.$$

We write the equation $\dot{x} = f(x)$ in the form

$$\begin{aligned} \dot{\varphi}_0(t) + \dot{\psi}(t) &= f_0(\varphi_0(t)) + Df_0(\varphi_0(t))\psi(t) \\ &\quad + [f(\varphi_0(t) + \psi(t)) - f_0(\varphi_0(t)) - Df_0(\varphi_0(t))\psi(t)]. \end{aligned} \tag{2.4}$$

First we consider the auxiliary linear equation

$$\dot{\chi}(t) = Df_0(\varphi_0(t))\chi(t) = ka\left(\frac{c}{t^p}\right)^{k-1}\chi(t) = \frac{-k}{k-1}\frac{1}{t}\chi(t)$$

which has a solution

$$\chi(t) = \frac{1}{t^{p+1}}.$$

From (2.4), using the variation of constants formula we get the following integral equation for ψ

$$\begin{aligned} \psi(t) &= \Gamma\psi(t) \\ &\equiv \frac{1}{t^{p+1}} \int_T^t s^{p+1}[f(\varphi_0(s) + \psi(s)) - f_0(\varphi_0(s)) - Df_0(\varphi_0(s))\psi(s)]\, ds. \end{aligned} \tag{2.5}$$

Here, we are implicitly assuming that $\psi(T) = 0$. We fix ν such that

$$p < \nu < \min\{q, p+1\}.$$

We introduce the space

$$X = \{\psi : \Omega(T, \alpha) \to \mathbb{C} : \text{analytic},\ |t|^\nu |\psi(t)| < \infty\}.$$

We endow X with the norm

$$\|\psi\| = \sup_{t\in\Omega(T,\alpha)} |t|^\nu |\psi(t)|.$$

We denote by $B(\rho)$ the closed ball of radius ρ in X centered at zero. We consider the operator $\Gamma : B(\rho) \to B(\rho)$ defined by (2.5).

The rest of this section is devoted to prove that, taking suitable values for ρ and T, we have that $\Gamma : B(\rho) \to B(\rho)$ is a contraction.

We take T satisfying the conditions

$$T \geq 1, \qquad T^p > 2c/r, \qquad T^{\nu-p} > 1/2,$$
$$\left(1+\frac{1}{2T^{\nu-p}}\right)^{k-2} \leq 2, \qquad \left(1+\frac{1}{2T^{\nu-p}}\right)^{\ell} \leq 2$$

and ρ such that

$$\begin{array}{ll} \rho < (c/2)\sin(\pi-\alpha) & \text{if } \ \pi/2 \leq \alpha < \pi \\ \rho < c/2 & \text{if } \ 0 \leq \alpha < \pi/2. \end{array}$$

First we check that Γ is well defined. If $\psi \in B(\rho)$,

$$\left|\frac{c}{s^p} + \psi(s)\right| \leq \frac{c}{T^p} + \frac{1}{T^\nu}\|\psi\| \leq \frac{1}{T^p}2c < r.$$

Moreover, for $s \in \Omega(T,\alpha)$,

$$|\arg\varphi_0(s)| = |\arg s^p| < p\alpha < \pi$$

and

$$\left|\frac{\psi(s)}{\varphi_0(s)}\right| = \left|\frac{\psi(s)}{c/s^p}\right| \leq \frac{\|\psi\|}{c}\frac{1}{|s|^{\nu-p}} \leq \frac{2\rho}{c}.$$

This implies that $|\arg(\varphi_0(s)+\psi(s))| < \pi$ and hence, if $\psi \in B(\rho)$, then for all $s \in \Omega(T,\alpha)$, $\varphi_0(s)+\psi(s)$ belongs to the domain of f.

We will use the following bounds. For $k \in \mathbb{R}$, $k > 1$ and $z, w \in \mathbb{C}$, such that z, $z+w \in U$ and $|w/z| < 1$, we have that

$$\begin{aligned} |(z+w)^k - z^k| &= |z|^k|(1+w/z)^k - 1| \leq |z|^k \sum_{j\geq 1}\left|\binom{k}{j}\right| |w/z|^j \\ &\leq |z|^{k-1}|w| \sum_{j\geq 1}\left|\binom{k}{j}\right| = C_{k,1}|z|^{k-1}|w| \end{aligned} \tag{2.6}$$

and

$$\begin{aligned} |(z+w)^k - z^k - kz^{k-1}w| &= |z|^k|(1+w/z)^k - 1 - k(w/z)| \\ &\leq |z|^k \sum_{j\geq 2}\left|\binom{k}{j}\right| |w/z|^j = C_{k,2}|z|^{k-2}|w|^2, \end{aligned}$$

where $C_{k,k_0} = \sum_{j\geq k_0}\left|\binom{k}{j}\right|$.

Given $t \in \Omega(T,\alpha)$, $t = |t|e^{i\theta}$, to evaluate the integral in the definition of Γ we will take the path of integration $\gamma = \gamma_1 \vee \gamma_2$ with $\gamma_1(u) = Te^{iu}$, $u \in [0,\theta]$, and $\gamma_2(u) = Te^{i\theta} + (t - Te^{i\theta})u$, $u \in [0,1]$. We call

$$I_\delta = \int_{\gamma_2} \frac{1}{|s|^\delta}\,ds = \frac{1}{e^{i\theta(\delta-1)}}\int_0^1 \frac{|t|-T}{[T+(|t|-T)u]^\delta}\,du.$$

We have $|I_\delta| = \frac{1}{1-\delta}(|t|^{1-\delta} - T^{1-\delta})$ if $\delta \neq 1$ and $|I_\delta| = \log\frac{|t|}{T}$ if $\delta = 1$.

We introduce

$$\chi_1(s) = s^{p+1}a\big[(\varphi_0(s)+\psi(s))^k - (\varphi_0(s))^k - k(\varphi_0(s))^{k-1}\psi(s)\big]$$

and $\chi_2(s) = s^{p+1}g(\varphi_0(s)+\psi(s))$. We have

$$\begin{aligned}
\left|\int_{\gamma_1}\chi_1(s)\,ds\right| &\le C_{k,2}\alpha|a|c^{k-2}\|\psi\|^2\frac{T}{T^{2(\nu-p)}},\\
\left|\int_{\gamma_1}\chi_2(s)\,ds\right| &\le 2Bc^\ell\alpha\frac{T}{T^{q-p}},\\
\left|\int_{\gamma_2}\chi_1(s)\,ds\right| &\le C_{k,2}|a|c^{k-2}\|\psi\|^2|I_{2(\nu-p)}|
\end{aligned}$$

and

$$\left|\int_{\gamma_2}\chi_2(s)\,ds\right| \le 2Bc^\ell|I_{q-p}|.$$

We have to distinguish the cases $2(\nu-p)<1$, $2(\nu-p)=1$, $2(\nu-p)>1$ and $q-p<1$, $q-p=1$ and $q-p>1$. Using the previous estimates we find that in all cases

$$|\Gamma(\psi(t))| \le \frac{\rho}{|t|^\nu}$$

if T is big enough, and hence $\Gamma\psi \in B(\rho)$.

Next we see that Γ is a contraction. Indeed, let ψ and $\tilde\psi$ be two functions which belong to $B(\rho)$,

$$\begin{aligned}
|(\Gamma\psi - \Gamma\tilde\psi)(t)| \le{}& \frac{1}{|t|^{p+1}}\left|\left(\int_{\gamma_1}+\int_{\gamma_2}\right)s^{p+1}\big(f_0(\varphi_0(s)+\psi(s)) - Df_0(\varphi_0(s))\psi(s)\right.\\
&-[f_0(\varphi_0(s)+\tilde\psi(s)) - Df_0(\varphi_0(s))\tilde\psi(s)]\big)\,ds \qquad (2.7)\\
&\left.+\left(\int_{\gamma_1}+\int_{\gamma_2}\right)s^{p+1}[g(\varphi_0(s)+\psi(s)) - g(\varphi_0(s)+\tilde\psi(s))]\,ds\right|.
\end{aligned}$$

To evaluate the first difference we consider the function

$$\chi(z) = a[(\varphi_0(s)+z)^k - k(\varphi_0(s))^{k-1}z].$$

By the mean value theorem, we have that

$$\chi(\tilde z) - \chi(z) = \int_0^1 \chi'(z+\zeta(\tilde z - z))[\tilde z - z]\,d\zeta$$

and, since $\chi'(w) = ak[(\varphi_0(s)+w)^{k-1} - (\varphi_0(s))^{k-1}]$ then, using (2.6),

$$|\chi(\tilde\psi(s)) - \chi(\psi(s))| \le |a|k\left[C_{k,1}\left(\frac{c}{|s|^p}\right)^{k-2}\frac{\rho}{|s|^{2\nu}}\|\tilde\psi - \psi\|\right].$$

We bound the first integrals in (2.7):

$$\begin{aligned}
\left|\int_{\gamma_1}s^{p+1}\big[\chi(\tilde\psi(s)) - \chi(\psi(s))\big]\,ds\right| &\le C_{k,1}k\alpha|a|c^{k-2}\rho\|\tilde\psi - \psi\|\frac{T}{T^{2(\nu-p)}},\\
\left|\int_{\gamma_2}s^{p+1}\big[\chi(\tilde\psi(s)) - \chi(\psi(s))\big]\,ds\right| &\le C_{k,1}k|a|c^{k-2}\rho\|\tilde\psi - \psi\|I_{2(\nu-p)}.
\end{aligned}$$

To deal with the second integrals we use that, since g is analytic in U, $|g'(z)| \leq B_2|z|^{\ell-1}$ in a domain

$$\{z \in \mathbb{C} : |z| < r_1, \, |\arg z| < \alpha_1\}, \quad 0 < r_1 < r, \;\; \alpha < \alpha_1 < \pi.$$

Then the integrals that we are analyzing are bounded by

$$\alpha B_2 c^{\ell-1}\Big(1 + \frac{\rho}{cT^{\nu-p}}\Big)^{\ell-1} \|\tilde{\psi} - \psi\| \frac{T}{T^{q-p+\nu-p}}$$

and

$$B_2 c^{\ell-1}\Big(1 + \frac{\rho}{cT^{\nu-p}}\Big)^{\ell-1} \|\tilde{\psi} - \psi\| I_{q-p+\nu-p}$$

respectively, where we have used that $p\ell = q + 1$. Distinguishing cases as before we get that

$$|(\Gamma\psi - \Gamma\tilde{\psi})(t)| \leq C_T \|\tilde{\psi} - \psi\| \frac{1}{|t|^\nu}$$

and C_T is smaller than 1 if T is big enough.

Hence, if T is big enough, Γ is a contraction in $B(\rho)$ and, by the fixed point theorem, there exists a unique solution of (2.5) which belongs to $B(\rho)$. This ends the proof of Proposition 2.1.

3. Parameterization of local invariant manifolds

3.1. Introduction

In this chapter we prove the existence of the local stable and the local unstable manifolds associated to the origin for the perturbed system. These are obtained through special parameterizations with properties which will be useful later on. We only deal with the local stable manifold, but it is clear that the results are also true for the unstable one, working with the system obtained reversing time.

In order to prove this result we need a good initial approximation of the stable (and unstable) manifold and suitable coordinates to work with.

In Section 3.3, we obtain these coordinates by canonical changes of variables using the averaging method. This method allows us to obtain two important things: remove the terms of order $\mu\varepsilon^p$ and remove the smallest degree terms (with respect to (x, y)) of h_1. We must average several times in order to obtain a high enough degree.

The initial approximation of the stable manifold is achieved as the invariant manifold of an appropriate intermediate system which is obtained by dropping the non-autonomous part in the averaged system. We remark that the construction of the initial approximation is only necessary when $k < n$. For $k \geq n$ we can use as initial approximation the unperturbed homoclinic orbit.

Finally we obtain a functional equation for the parameterization of the stable manifold and we prove it has a solution applying the fixed point theorem in a suitable Banach space.

It is important to say that although the system is C^0 in t/ε, we obtain a parameterization with two parameters, say (t, s), which is analytic in s, considered as a complex variable, and we provide a dynamic sense for s.

3.2. Definitions and main result

We begin by introducing some notation. Given $T > 0$ and $\tau > 0$ we define the following sets:

$$D^{\mathrm{s}} = D^{\mathrm{s}}(T, \tau) = \{(t, s) \in \mathbb{R} \times \mathbb{C} : \ t + \operatorname{Re} s \geq T, \ |\operatorname{Im} s| \leq \tau\}$$

and

$$D^{\mathrm{u}} = D^{\mathrm{u}}(T, \tau) = \{(t, s) \in \mathbb{R} \times \mathbb{C} : \ t + \operatorname{Re} s \leq -T, \ |\operatorname{Im} s| \leq \tau\}.$$

Note that if $(t, s) \in D^{\mathrm{s}}$, $(t + 2\pi\varepsilon, s)$ and $(t, s + 2\pi\varepsilon)$ also belong to D^{s}.

For $k \in \mathbb{R}$, $k \geq 0$, we define the space $\mathcal{X}_k^{\mathrm{s}}$ of functions $h : D^{\mathrm{s}} \to \mathbb{C}$ such that

(a) h is continuous.
(b) For t fixed, $s \mapsto h(t, s)$ is analytic in $|\operatorname{Re} s| \leq T - t$, $|\operatorname{Im} s| \leq \tau$.
(c) $h(t + 2\pi\varepsilon, s) = h(t, s + 2\pi\varepsilon)$ for all $(t, s) \in D^{\mathrm{s}}$.
(d) $\|h\|_k \equiv \sup\{(t + \operatorname{Re} s)^k |h(t, s)| : (t, s) \in D^{\mathrm{s}}\} < \infty$.

In the obvious analogous way we define $\mathcal{X}_k^{\mathrm{u}}$.

We endow $\mathcal{X}_k^{\mathrm{s}}$ with the norm $\|\cdot\|_k$ introduced in (d) and it becomes a Banach space. Clearly we have $\mathcal{X}_{k+1}^{\mathrm{s}} \subset \mathcal{X}_k^{\mathrm{s}}$.

The main theorem of this chapter states the existence of invariant manifolds for the perturbed Hamiltonian equation

$$\begin{aligned} \dot{x} &= y + \mu\varepsilon^p \partial_y h_1(x, y, t/\varepsilon, \mu, \varepsilon) \\ \dot{y} &= -V'(x) - \mu\varepsilon^p \partial_x h_1(x, y, t/\varepsilon, \mu, \varepsilon) \end{aligned} \tag{3.1}$$

assuming conditions on the orders of h_0 and h_1. It is worth noting that here, in contrast with hypothesis **HP3**, we do not assume that h_1 is a polynomial.

THEOREM 3.1. *Let*

$$H = h_0 + \mu\varepsilon^p h_1$$

where $h_0(x, y) = y^2/2 + V(x)$ and V is analytic, $V(x) = a_n x^n + \dots$, $a_n < 0$, $h_1 = h_1(x, y, t/\varepsilon, \mu, \varepsilon)$ is continuous, analytic in (x, y, μ) in a neighborhood of $(0, 0, 0)$, of order k in (x, y), 2π-periodic in t/ε with zero average. We assume that $2k - 2 \geq n$, and in case that $p < 1$, $\partial_x h_1 \partial_y h_1 = 0$.

Then, H has stable and unstable invariants manifolds associated to $(0, 0)$, and, given $\tau > 0$, there exist $T > 0$ big enough and parameterizations $\gamma^{\mathrm{s}}_{\mu,\varepsilon}(t, s)$, $\gamma^{\mathrm{u}}_{\mu,\varepsilon}(t, s)$ of the local stable and unstable invariant manifolds, defined in $D^{\mathrm{s}}(T, \tau)$, $D^{\mathrm{u}}(T, \tau)$, respectively, such that ($*$ stands for s or u):

1) *$t \mapsto \gamma^*_{\mu,\varepsilon}(t, s)$ is a solution of system (3.1) and $s \mapsto \gamma^*_{\mu,\varepsilon}(t, s)$ is real analytic. Moreover the map $(t, s, \mu, \varepsilon) \mapsto \gamma^*_{\mu,\varepsilon}(t, s)$ is continuous, C^1 with respect to t and analytic with respect to (s, μ).*
2) *For all $(t, s) \in D^*(T, \tau)$*

$$\gamma^*_{\mu,\varepsilon}(t \pm 2\pi\varepsilon, s) = \gamma^*_{\mu,\varepsilon}(t, s \pm 2\pi\varepsilon)$$

where we take $+$ for $* = \mathrm{s}$ and $-$ for $* = \mathrm{u}$.

3) *For $\mu = 0$, $\gamma^*_{\mu,\varepsilon}(t, s)$ coincides with the restriction of the homoclinic solution $\gamma_0(t + s)$ to $D^*(T, \tau)$, and for $\mu \neq 0$ the following estimate holds:*

$$\gamma^*_{\mu,\varepsilon}(t, s) = \gamma_0(t + s) + \mu\varepsilon^{p+1}\sigma^*_{\mu,\varepsilon}(t, s)$$

where $\sigma^*_{\mu,\varepsilon}(t, s) \in \mathcal{X}^*_\lambda \times \mathcal{X}^*_\lambda$ with $\lambda = \frac{2}{n-2}$ and

$$\sigma^*_{\mu,\varepsilon}(t, s) = G_{\mu,\varepsilon}(\gamma_0(t + s), t/\varepsilon) + O(\varepsilon)$$

where $G_{\mu,\varepsilon} = (G_1, G_2)$ is determined by the conditions

$$\partial_\theta G_{\mu,\varepsilon}(x, y, \theta) = (\partial_y h_1(x, y, \theta, \mu, \varepsilon), -\partial_x h_1(x, y, \theta, \mu, \varepsilon)),$$

and having zero mean.

REMARK 3.1. *In this theorem we have introduced a new condition: if $p < 1$ then $\partial_x h_1 \partial_y h_1 = 0$. By Remark 1.5, hypothesis **HP5** implies this condition, and therefore, under hypotheses **HP1-HP5**, Theorem 3.1 applies.*

The proof of this theorem is done in several steps in the present chapter. We prove the statements for $\gamma^{\mathrm{s}}_{\mu,\varepsilon}$, the ones for $\gamma^{\mathrm{u}}_{\mu,\varepsilon}$ are obtained changing t by $-t$.

In the following sections we assume the hypotheses of Theorem 3.1. From now on, to simplify the notation, we omit the dependence on ε and μ at several places where it does not play an essential role.

3.3. Averaging of the equation

Some steps of averaging are necessary to transform equation (3.1) into a suitable form. We begin by introducing notation.

DEFINITION 3.1. *Given $l \in \mathbb{Z}^+$ and $U \subset \mathbb{C}^2$ we denote by P_l the set of functions $p : U \times \mathbb{R} \times B(0, \mu_0) \times [0, \varepsilon_0) \to \mathbb{C}$ that are continuous, 2π-periodic in θ, analytic in (x, y, μ), and have order l. Therefore they can be represented in the form*

$$p_{\mu,\varepsilon}(x, y, \theta) = p(x, y, \theta, \mu, \varepsilon) = \sum_{i+j=l}^{\infty} a_{i,j}(\theta, \mu, \varepsilon)x^i y^j$$

with the coefficients $a_{i,j}(\theta, \mu, \varepsilon)$ continuous, 2π-periodic in θ, and analytic in μ.

We will write $p = (p_{l_1}, p_{l_2}) \in P_{l_1} \times P_{l_2}$ if $p_{l_1} \in P_{l_1}$ and $p_{l_2} \in P_{l_2}$. However, if $l_1 = l_2$ we will simply write that $p \in P_{l_1}$.

For notational convenience we also define

$$P_l = P_0 \qquad \text{for} \qquad l < 0.$$

PROPOSITION 3.1. *There exists a canonical change of variables $\mathcal{C}$ defined in a neighborhood of the origin, that transforms the Hamiltonian H to*

$$\begin{aligned} \mathcal{H}(\bar{x}, \bar{y}, t/\varepsilon, \mu, \varepsilon) &= h_0(\bar{x}, \bar{y}) + \mu\varepsilon^{p+3} F_{2n-2}(\bar{x}, \bar{y}, t/\varepsilon, \mu, \varepsilon) \\ &\quad + \mu^2 \varepsilon^{p+2} R_{2k-2}(\bar{x}, \bar{y}, \mu, \varepsilon) \end{aligned}$$

in a neighborhood of the origin, where $F_{2n-2} \in P_{2n-2}$ and has zero mean, $R_{2k-2} \in P_{2k-2}$ does not depend on t and

$$R_{2k-2} = \overline{\partial_y h_1 \partial_x S} + \varepsilon r_{2k-2}$$

with $S \in P_k$ such that $\partial_\theta S(x, y, \theta, \mu, \varepsilon) = -h_1(x, y, \theta, \mu, \varepsilon)$ and has zero mean, and $r_{2k-2} \in P_{2k-2}$. The change $\mathcal{C}$ has the form

$$\mathcal{C}(\bar{x}, \bar{y}, t/\varepsilon, \mu, \varepsilon) = (\bar{x}, \bar{y}) + \mu\varepsilon^{p+1} G(\bar{x}, \bar{y}, t/\varepsilon, \mu, \varepsilon) + O(\mu\varepsilon^{p+2})$$

where G is determined by $\partial_\theta G = (\partial_y h_1, -\partial_x h_1)$ and having zero mean.

Moreover $\mathcal{C}$ and $\mathcal{H}$ are continuous in $(\bar{x}, \bar{y}, \theta = t/\varepsilon, \mu, \varepsilon)$, 2π-periodic in θ, and analytic in $(\bar{x}, \bar{y}, \mu)$ and $\mathcal{C}$ is C^1 in θ.

To average first we scale time by $\theta = t/\varepsilon$. The Hamiltonian becomes εH and the corresponding equation is

$$\begin{aligned} \dot{x} &= \varepsilon y + \mu\varepsilon^{p+1} \partial_y h_1(x, y, \theta, \mu, \varepsilon) \\ \dot{y} &= -\varepsilon V'(x) - \mu\varepsilon^{p+1} \partial_x h_1(x, y, \theta, \mu, \varepsilon) \end{aligned}$$

where $\dot{x}$ and $\dot{y}$ now mean derivatives with respect to the new time θ. Next we average several times with respect to θ in order to move the contribution of the perturbation to terms of order $\mu\varepsilon^{p+3}$ and $\mu^2\varepsilon^{p+3}$ in the parameters, and also to increase the orders with respect to x, y of some terms of H. We will make two sets of averaging steps. We begin with the first set.

Let $\nu \in \mathbb{Z}^+$. For the inductive step of averaging we assume we have a Hamiltonian

$$\varepsilon \mathcal{H}^\nu = \varepsilon h_0 + \mu\varepsilon^{p+\nu+1} F^\nu + \mu^2 \varepsilon^{2p+2} R^\nu_{2k-2}$$

with $R^\nu_{2k-2} \in P_{2k-2}$ and F^ν having the form

$$F^\nu = \sum_{\substack{l\in\mathbb{Z}^+ \\ \nu-2l\geq 0}} y^{\nu-2l} p_{i_l^\nu} + \sum_{\substack{l\in\mathbb{Z}^+ \\ \nu-2l-1\geq 0}} y^{\nu-2l-1} p_{j_l^\nu} \tag{3.2}$$

with $i_l^\nu = \max\{0, k+nl-\nu\}$, $j_l^\nu = \max\{0, k+n(l+1)-\nu-1\}$ and $p_i \in P_i$. We assume that F^ν has zero mean. From (3.2) it is clear that $F^\nu \in P_k$. Indeed, the terms are of order either

$$\nu - 2l + i_l^\nu \geq \nu - 2l + k + nl - \nu = k + l(n-2) \geq k$$

or

$$\nu - 2l - 1 + j_l^\nu \geq \nu - 2l - 1 + k + n(l+1) - \nu - 1 = k + n - 2 + l(n-2) \geq k.$$

We observe that the original Hamiltonian εH has this form for $\nu = 0$ with $F^0 = h_1$ and $R^0_{2k-2} = 0$.

LEMMA 3.1. *Under the previous conditions and assuming that $n \geq 3$ and $k \geq 2$, there exists a canonical change of variables $(x,y) = \mathcal{C}^{\nu+1}(\bar{x},\bar{y},\theta,\mu,\varepsilon)$ which is C^0 in $(\bar{x},\bar{y},\theta,\mu,\varepsilon)$, C^1 and 2π-periodic in θ and analytic in $(\bar{x},\bar{y},\mu)$ that transforms the Hamiltonian $\varepsilon\mathcal{H}^\nu$ to*

$$\varepsilon\mathcal{H}^{\nu+1} = \varepsilon h_0 + \mu\varepsilon^{p+\nu+2} F^{\nu+1} + \mu^2 \varepsilon^{2p+2} R^{\nu+1}_{2k-2}$$

in a neighborhood of the origin, where

$$F^{\nu+1} = \bar{y}\partial_x S_1^{\nu+1} - V'(\bar{x})\partial_{\bar{y}} S_1^{\nu+1}$$

and $S_1^{\nu+1} = S_1^{\nu+1}(\bar{x},\bar{y},\theta,\mu,\varepsilon)$ satisfies

$$\partial_\theta S_1^{\nu+1} = -F^\nu, \tag{3.3}$$

and has zero mean.

Moreover $F^{\nu+1}$ has the form (3.2), $F^{\nu+1} \in P_k$ and has zero mean with respect to θ, $R^{\nu+1}_{2k-2} \in P_{2k-2}$ and

$$R^{\nu+1}_{2k-2} = \varepsilon^{2\nu}\partial_y F^\nu \partial_x S_1^{\nu+1} + R^\nu_{2k-2} + \varepsilon^{2\nu+1} r_{2k-2} \tag{3.4}$$

with $r_{2k-2} \in P_{2k-2}$.

REMARK 3.2. *To prove this lemma we do not assume the condition $2k-2 \geq n$.*

PROOF. To simplify the notation, in this proof we will not write the dependence of the functions on μ, ε. Also, along the proof r_j and g_j will mean generic terms of P_j; therefore they may be different at different places. We consider a generating function $S^{\nu+1}(x,\bar{y},\theta)$ which will provide a canonical change of variables $(\bar{x},\bar{y}) \mapsto (x,y)$ implicitly through

$$\begin{aligned} \bar{x} &= \partial_{\bar{y}} S^{\nu+1}(x,\bar{y},\theta) \\ y &= \partial_x S^{\nu+1}(x,\bar{y},\theta) \end{aligned} \tag{3.5}$$

and then the new Hamiltonian will be

$$\varepsilon\mathcal{H}^{\nu+1}(\bar{x},\bar{y},\theta) = \varepsilon\mathcal{H}^\nu(x,y,\theta) + \partial_\theta S^{\nu+1}(x,\bar{y},\theta).$$

We take

$$S^{\nu+1}(x,\bar{y},\theta) = x\bar{y} + \mu\varepsilon^{p+\nu+1} S_1^{\nu+1}(x,\bar{y},\theta)$$

with $S_1^{\nu+1}$ satisfying (3.3). This choice is motivated by the next calculations. We observe that $S^{\nu+1}$ is C^0, 2π-periodic in θ, C^1 in (x, y, θ, μ) and analytic in (x, y, μ). With it we will cancel the terms of orders $\mu\varepsilon^{p+\nu+1}$ in the averaged Hamiltonian. Since F^ν has zero mean with respect to θ we can choose $S_1^{\nu+1}$ with the additional condition of having zero mean.

From (3.3), $S_1^{\nu+1}$ is of order k. To get the change of variables, from (3.5) we have to apply the implicit function theorem (I.F.T.) to

$$(x, y, \bar{x}, \bar{y}, \theta, \mu, \varepsilon) \mapsto (\bar{x} - \partial_{\bar{y}} S^{\nu+1}(x, \bar{y}, \theta, \mu, \varepsilon), y - \partial_x S^{\nu+1}(x, \bar{y}, \theta, \mu, \varepsilon)).$$

This map is C^1 with respect to $(x, y, \bar{x}, \bar{y}, \theta, \mu)$ and continuous. A generalized version of the I.F.T. gives that we can obtain

$$(x, y) = \mathcal{C}^{\nu+1}(\bar{x}, \bar{y}, \theta, \mu, \varepsilon)$$

with $\mathcal{C}^{\nu+1}$ being C^1 with respect to x, y, θ, μ and continuous.

A new application of the I.F.T. for analytic functions, with θ and ε fixed, gives, by uniqueness, that $C^{\nu+1}$ also is analytic with respect to (x, y, μ).

Moreover it is clear that

$$\begin{aligned} x &= \bar{x} - \mu\varepsilon^{p+\nu+1}\partial_{\bar{y}} S_1^{\nu+1} + \mu^2\varepsilon^{2p+2\nu+2} r_{2k-3} \\ y &= \bar{y} + \mu\varepsilon^{p+\nu+1}\partial_x S_1^{\nu+1} + \mu^2\varepsilon^{2p+2\nu+2} r_{2k-3} \end{aligned}$$

where the derivatives of $S_1^{\nu+1}$ and r_{2k-3} are evaluated at $(\bar{x}, \bar{y}, \theta)$. The averaged Hamiltonian is

$$\begin{aligned} \varepsilon\mathcal{H}^{\nu+1}(\bar{x}, \bar{y}, \theta) &= \varepsilon\mathcal{H}^\nu(x, y, \theta) - \mu\varepsilon^{p+\nu+1} F^\nu(x, \bar{y}, \theta) \\ &= \frac{\varepsilon}{2}[\bar{y} + \mu\varepsilon^{p+\nu+1}\partial_x S_1^{\nu+1} + \mu^2\varepsilon^{2p+2\nu+2} r_{2k-3}]^2 \\ &\quad +\varepsilon V(\bar{x} - \mu\varepsilon^{p+\nu+1}\partial_{\bar{y}} S_1^{\nu+1} + \mu^2\varepsilon^{2p+2\nu+2} r_{2k-3}) \\ &\quad +\mu\varepsilon^{p+\nu+1}[F^\nu(x, y, \theta) - F^\nu(x, \bar{y}, \theta)] + \mu^2\varepsilon^{2p+2} R_{2k-2}^\nu \\ &= \frac{\varepsilon}{2}\bar{y}^2 + \varepsilon V(\bar{x}) + \mu\varepsilon^{p+\nu+2}[\bar{y}\partial_x S_1^{\nu+1} - V'(\bar{x})\partial_{\bar{y}} S_1^{\nu+1}] \\ &\quad +\mu^2\varepsilon^{2p+2\nu+2}\partial_y F^\nu \partial_x S_1^{\nu+1} + \mu^2\varepsilon^{2p+2} R_{2k-2}^\nu + \mu^2\varepsilon^{2p+2\nu+3} r_{2k-2}. \end{aligned}$$

Therefore we can take

$$R_{2k-2}^{\nu+1} = \varepsilon^{2\nu}\partial_y F^\nu \partial_x S_1^{\nu+1} + R_{2k-2}^\nu + \varepsilon^{2\nu+1} r_{2k-2}. \tag{3.6}$$

We will need information on the orders of the terms in the Hamiltonian and the factors y they have. From (3.2) and (3.3) we write

$$S_1^{\nu+1} = \sum_{\nu-2l\geq 0} y^{\nu-2l} g_{i_l^\nu} + \sum_{\nu-2l-1\geq 0} y^{\nu-2l-1} g_{j_l^\nu}$$

when here and in what follows l is a non-negative integer.

Therefore, $F^{\nu+1} = y\partial_x S_1^{\nu+1} - V'(x)\partial_y S_1^{\nu+1}$ is of the form

$$\begin{aligned} F^{\nu+1} &= \sum_{\nu-2l\geq 0} y^{\nu+1-2l} g_{\max\{0, i_l^\nu - 1\}} + \sum_{\nu-2l-1\geq 0} y^{\nu+1-2l-1} g_{\max\{0, j_l^\nu - 1\}} \\ &\quad + \sum_{\nu-2l\geq 0} y^{\nu+1-2(l+1)} g_{i_l^\nu + n - 1} + \sum_{\nu-2l\geq 0} y^{\nu+1-2l-1} g_{i_l^\nu + n - 2} \\ &\quad + \sum_{\nu-2l-1\geq 0} y^{\nu+1-2(l+1)-1} g_{j_l^\nu + n - 1} + \sum_{\nu-2l-1\geq 0} y^{\nu+1-2(l+1)} g_{j_l^\nu + n - 2}. \end{aligned}$$

The following relations hold

$$\begin{aligned}\max\{0, i_l^\nu - 1\} &\geq i_l^{\nu+1}, \qquad \max\{0, j_l^\nu - 1\} \geq j_l^{\nu+1},\\ i_l^\nu + n - 1 &\geq i_{l+1}^{\nu+1}, \qquad j_l^\nu + n - 1 \geq j_{l+1}^{\nu+1},\\ i_l^\nu + n - 1 &\geq i_{l+1}^{\nu+1}, \qquad j_l^\nu + n - 1 \geq j_{l+1}^{\nu+1},\end{aligned}$$

and then we deduce that $F^{\nu+1}$ has the form (3.2).

Since $S_1^{\nu+1}$ has zero mean, we also have that $F^{\nu+1}$ has zero mean.

Finally the regularity on $\mathcal{H}^{\nu+1}$ follows from the regularity of $\mathcal{H}^\nu$ and $\mathcal{C}^{\nu+1}$. □

Now we use the previous lemma to perform $2n+2$ steps of averaging.

LEMMA 3.2. *There exists a canonical change of variables* $(x, y) = \mathcal{C}_0(\bar{x}, \bar{y}, \theta, \mu, \varepsilon)$ *which is* C^0 *in* $(\bar{x}, \bar{y}, \theta, \mu, \varepsilon)$, C^1 *and* 2π*-periodic in* θ *and analytic in* $(\bar{x}, \bar{y}, \mu)$ *that transforms the Hamiltonian* εH *to*

$$\varepsilon\mathcal{H}_0 = \varepsilon h_0 + \mu\varepsilon^{p+2n+3}F + \mu^2\varepsilon^{2p+2}R_{2k-2}$$

in a neighborhood of the origin, where $F \in P_{2n-2}$ *and has zero mean with respect to* θ, $R_{2k-2} \in P_{2k-2}$ *and*

$$R_{2k-2} = \partial_y h_1 \partial_x S_1^1 + \varepsilon r_{2k-2}$$

with S_1^1 *such that* $\partial_\theta S_1^1 = -h_1$ *and has zero mean, and* $r_{2k-2} \in P_{2k-2}$. *Moreover* $\mathcal{H}_0$ *is continuous in* $(\bar{x}, \bar{y}, \theta, \mu, \varepsilon)$ *and analytic in* $(\bar{x}, \bar{y}, \mu)$.

PROOF. Since $h_1 \in P_k$, we note that H has the form (3.2) for $\nu = 0$ and $R_{2k-2}^0 = 0$. Then we begin with $F^0 = h_1$ and $R_{2k-2}^0 = 0$ and we apply iteratively Lemma 3.1 $2n+2$ times. In this way we obtain that $F \equiv F^{2n+2}$ has the form

$$F^{2n+2} = \sum_{0 \leq l \leq n+1} \bar{y}^{2n+2-2l} p_{i_l} + \sum_{0 \leq 2l \leq 2n+1} \bar{y}^{2n+1-2l} p_{j_l} \tag{3.7}$$

with $i_l = \max\{0, k + nl - 2n - 2\}$, $j_l = \max\{0, k + n(l+1) - 2n - 3\}$ and $p_m \in P_m$. Also from (3.6), we can write R_{2k-2}^{2n+2} as

$$R_{2k-2}^{2n+2} = \partial_y h_1 \partial_x S_1^1 + \varepsilon r_{2k-2}$$

where S_1^1 is the one which corresponds to the first change $\mathcal{C}^1$. Moreover the function F^{2n+2} has zero mean with respect to θ. We observe that the Hamiltonians $\mathcal{H}^1, \dots, \mathcal{H}^{2n+2}$ are C^0, 2π-periodic in θ and analytic with respect to x, y and μ. The changes $\mathcal{C}^1, \dots, \mathcal{C}^{2n+2}$ are C^1 with respect to θ. We take $\mathcal{C}_0 = \mathcal{C}^{2n+2} \circ \cdots \circ \mathcal{C}^1$.

We prove now that, if a function has the form given in (3.7), then it belongs to P_{2n-2}. For that we check the order of every term in (3.7). In $\sum_{l \leq n+1} \bar{y}^{2n+2-2l} p_{i_l}$, the term indexed by l has order $2n + 2 - 2l + i_l \geq 2n$ if $l = 0, 1$; and

$$2n + 2 - 2l + i_l \geq (n-2)l + k \geq 2(n-1) + k - 2,$$

if $l \geq 2$. In $\sum_{2l \leq 2n+1} \bar{y}^{2n+1-2l} p_{j_l}$, it has order bigger than $2n+1$ if $l = 0$ and

$$2n + 1 - 2l + j_l \geq k + (n-2)(l+1) \geq 2(n-1) + k - 2$$

if $l \geq 1$. Hence $F = F^{2n+2} \in P_{2n-2}$. □

REMARK 3.3. *We observe that* $\mu^2\varepsilon^{2p+2}R_{2k-2}$ *can be written as* $\mu^2\varepsilon^{p+3}R_{2k-2}$. *Indeed, if* $p \geq 1$ *it follows from the comparison of powers of* ε. *And, if* $p < 1$, *by hypothesis* $\partial_y h_1 \partial_x h_1 = 0$ *which implies that* $\partial_y h_1 \partial_x S_1^1 = 0$.

We rename the variables $(\bar{x}, \bar{y})$ by (x, y). According to the previous remark, the transformed Hamiltonian is

$$\varepsilon\mathcal{H}_0 \quad = \quad \varepsilon h_0 + \mu\varepsilon^{p+2n+3}F_{2n-2} + \mu^2\varepsilon^{p+3}R_{2k-2}.$$

Next we perform a new set of steps of averaging in order to eliminate the dependence on θ in the terms of R_{2k-2} of order (in the (x, y) variables) smaller than $2n-2$.

As usual, given $f \in P_j$, we will denote its mean with respect to θ by $\bar{f}$. We proceed in a similar way as before. For the inductive step, we consider a Hamiltonian system of the form

$$\begin{aligned}\varepsilon\tilde{\mathcal{H}}^\nu(x, y, \theta, \mu, \varepsilon) \quad &= \quad \varepsilon h_0(x, y) + \mu^2\varepsilon^{p+3}R^\nu_{2k-2}(x, y, \mu, \varepsilon) \\ &\quad +\mu\varepsilon^{p+4}F^\nu_{2n-2}(x, y, \theta, \mu, \varepsilon) + \mu^2\varepsilon^{p+3}T^\nu(x, y, \theta, \mu, \varepsilon)\end{aligned}$$

where $F^\nu_{2n-2} \in P_{2n-2}$, $R^\nu_{2k-2} \in P_{2k-2}$, T^ν has zero mean and has the form

$$T^\nu = y^\nu r_{i_\nu} \tag{3.8}$$

with $i_\nu = \max\{0, 2k-2-\nu\}$ and $r_{i_\nu} \in P_{i_\nu}$.

We observe that $\varepsilon\mathcal{H}_0$ has this form for $\nu = 0$, with $F^0_{2n-2} = \varepsilon^{2n-1}F_{2n-2}$, R^0_{2k-2} the mean of R_{2k-2} and $T^0 = R_{2k-2} - R^0_{2k-2}$.

LEMMA 3.3. *Under the previous conditions and assuming that $n \geq 3$ and $2k-2 \geq n$, there exists a canonical change of variables $(x, y) = \tilde{C}^{\nu+1}(\bar{x}, \bar{y}, \theta, \mu, \varepsilon)$ which is C^0, C^1 and 2π-periodic in θ and analytic in $(\bar{x}, \bar{y}, \mu)$ that transforms the Hamiltonian $\varepsilon\tilde{\mathcal{H}}^\nu$ to*

$$\begin{aligned}\varepsilon\tilde{\mathcal{H}}^{\nu+1}(\bar{x}, \bar{y}, \theta, \mu, \varepsilon) \quad &= \quad \varepsilon h_0(\bar{x}, \bar{y}) + \mu^2\varepsilon^{p+3}R^{\nu+1}_{2k-2}(\bar{x}, \bar{y}, \mu, \varepsilon) \\ &\quad +\mu\varepsilon^{p+4}F^{\nu+1}_{2n-2}(\bar{x}, \bar{y}, \theta, \mu, \varepsilon) + \mu^2\varepsilon^{p+3}T^{\nu+1}(\bar{x}, \bar{y}, \theta, \mu, \varepsilon)\end{aligned}$$

in a neighborhood of the origin, where

$$\begin{aligned}T^{\nu+1} \quad &= \quad \varepsilon\bar{y}\partial_x S^{\nu+1}_1 \\ R^{\nu+1}_{2k-2} \quad &= \quad R^\nu_{2k-2} + \varepsilon r_{2k-2} \\ F^{\nu+1}_{2n-2} \quad &= \quad F^\nu_{2n-2} + \mu(r_{2n-2} - \bar{r}_{2n-2}).\end{aligned}$$

The function $S^{\nu+1}_1(\bar{x}, \bar{y}, \theta, \mu, \varepsilon)$ is determined by the conditions

$$\partial_\theta S^{\nu+1}_1 = -T^\nu, \qquad \overline{S^{\nu+1}_1} = 0. \tag{3.9}$$

Moreover

$$T^{\nu+1} = y^{\nu+1}p_{i_{\nu+1}},$$

with $p_j \in P_j$, and $T^{\nu+1}$ has zero mean. Also $\tilde{\mathcal{H}}^{\nu+1}$ is continuous and analytic in $(\bar{x}, \bar{y}, \mu)$.

PROOF. In the proof we will not write the dependence of the functions on μ, ε. Along the proof r_j will mean a generic term of P_j. As before, we consider a generating function $S^{\nu+1}(x, \bar{y}, \theta)$ which will provide a canonical change of variables $(\bar{x}, \bar{y}) \mapsto (x, y)$ implicitly through

$$\begin{aligned}\bar{x} \quad &= \quad \partial_{\bar{y}}S^{\nu+1}(x, \bar{y}, \theta) \\ y \quad &= \quad \partial_x S^{\nu+1}(x, \bar{y}, \theta).\end{aligned} \tag{3.10}$$

We take

$$S^{\nu+1}(x,\bar{y},\theta) = x\bar{y} + \mu^2\varepsilon^{p+3}S_1^{\nu+1}(x,\bar{y},\theta)$$

with $S_1^{\nu+1}$ satisfying (3.9). We observe that $S^{\nu+1}$ is C^0, $C^1$and 2π-periodic in θ and analytic in (x,y,μ).

From (3.8), it is clear that $T^\nu \in P_{2k-2}$. Hence $S_1^{\nu+1} \in P_{2k-2}$.

Analogously as in the proof of Lemma 3.1 we get the regularity statements for $\tilde{\mathcal{C}}^{\nu+1}$ and we have that

$$\begin{aligned} x &= \bar{x} - \mu^2\varepsilon^{p+3}\partial_{\bar{y}}S_1^{\nu+1} + \mu^4\varepsilon^{2p+6}r_{4k-7} \\ y &= \bar{y} + \mu^2\varepsilon^{p+3}\partial_x S_1^{\nu+1} + \mu^4\varepsilon^{2p+6}r_{4k-7} \end{aligned}$$

where the derivatives of $S_1^{\nu+1}$ and r_{4k-7} are evaluated at $(\bar{x},\bar{y},\theta)$. The averaged Hamiltonian is therefore

$$\begin{aligned} \varepsilon\tilde{\mathcal{H}}^{\nu+1}(\bar{x},\bar{y},\theta) &= \varepsilon\tilde{\mathcal{H}}^{\nu}(x,y,\theta) - \mu^2\varepsilon^{p+3}T^\nu(x,\bar{y},\theta) \\ &= \frac{\varepsilon}{2}[\bar{y} + \mu^2\varepsilon^{p+3}\partial_x S_1^{\nu+1} + \mu^4\varepsilon^{2p+6}r_{4k-7}]^2 \\ &\quad +\varepsilon V(\bar{x} - \mu^2\varepsilon^{p+3}\partial_{\bar{y}}S_1^{\nu+1} + \mu^4\varepsilon^{2p+6}r_{4k-7}) \\ &\quad +\mu^2\varepsilon^{p+3}[T^\nu(x,y,\theta) - T^\nu(x,\bar{y},\theta)] \\ &\quad +\mu^2\varepsilon^{p+3}R_{2k-2}^\nu + \mu\varepsilon^{p+4}F_{2n-2}^\nu + \mu^4\varepsilon^{2p+6}r_{4k-6} \\ &= \frac{\varepsilon}{2}\bar{y}^2 + \varepsilon V(\bar{x}) + \mu^2\varepsilon^{p+3}R_{2k-2}^\nu + \mu^2\varepsilon^{p+4}\bar{y}\partial_x S_1^{\nu+1} \\ &\quad +\mu\varepsilon^{p+4}F_{2n-2}^\nu + \mu^2\varepsilon^{p+4}r_{2n-2} \end{aligned}$$

where in the last equality we have used that, since $2k-2 \geq n$, we have that $4k-6 \geq 2n-2$.

Therefore we can take

$$\begin{aligned} F_{2n-2}^{\nu+1} &= F_{2n-2}^\nu + \mu(r_{2n-2} - \bar{r}_{2n-2}) \\ R_{2k-2}^{\nu+1} &= R_{2k-2}^\nu + \varepsilon\bar{r}_{2n-2} \\ T^{\nu+1} &= \varepsilon\bar{y}\partial_x S_1^{\nu+1}. \end{aligned}$$

Now we check that $T^{\nu+1}$ has the claimed form. It is clear that it has zero mean, and that, from (3.8) and (3.9), $T^{\nu+1}$ has the form

$$T^{\nu+1} = y^{\nu+1}p_{\max\{0,i_\nu-1\}}$$

and $i_{\nu+1} = \max\{0, i_\nu - 1\}$.

Finally, the regularity statement on $\tilde{\mathcal{H}}^{\nu+1}$ follows from the regularity of $\tilde{\mathcal{H}}^\nu$ and $\tilde{\mathcal{C}}^{\nu+1}$, in the same way as in Lemma 3.1. □

Now we use the previous lemma to perform $2n-2$ steps of averaging.

LEMMA 3.4. *There exists a canonical change of variables $(x,y) = \mathcal{C}_1(\bar{x},\bar{y},\theta,\mu,\varepsilon)$ which is C^0, C^1 and 2π-periodic in θ and analytic in $(\bar{x},\bar{y},\mu)$ that transforms the Hamiltonian $\varepsilon\mathcal{H}_0$ to*

$$\begin{aligned} \varepsilon\mathcal{H}_1(\bar{x},\bar{y},\theta,\mu,\varepsilon) &= \varepsilon h_0(\bar{x},\bar{y}) + \mu\varepsilon^{p+4}F_{2n-2}(\bar{x},\bar{y},\theta,\mu,\varepsilon) \\ &\quad +\mu^2\varepsilon^{p+3}R_{2k-2}(\bar{x},\bar{y},\mu,\varepsilon) \end{aligned}$$

in a neighborhood of the origin, where $F_{2n-2} \in P_{2n-2}$ *and has zero mean,* $R_{2k-2} \in P_{2k-2}$ *does not depend on* θ *and*

$$R_{2k-2} = \overline{\partial_y h_1 \partial_x S_1^1} + \varepsilon r_{2k-2}$$

with S_1^1 *such that* $\partial_\theta S_1^1 = -h_1$ *and has zero mean, and* $r_{2k-2} \in P_{2k-2}$. *Moreover* $\mathcal{H}_1$ *is continuous, and analytic in* $(\bar{x}, \bar{y}, \mu)$.

PROOF. We apply Lemma 3.3 iteratively $2n-2$ times. We omit the details since they are the same as the ones in Lemma 3.2. □

PROOF OF PROPOSITION 3.1. First we scale time to obtain the Hamiltonian εH. Then we apply the changes $\mathcal{C}_0$ and $\mathcal{C}_1$ given by Lemmas 3.2 and 3.4 respectively to obtain the Hamiltonian $\varepsilon\mathcal{H}_1$, and finally we scale back to the original time. We take $\mathcal{H} = \mathcal{H}_1(x, y, t/\varepsilon, \mu, \varepsilon)$. To obtain the form of $\mathcal{C}$ just recall that $\mathcal{C} = \mathcal{C}_1 \circ \mathcal{C}_0$, $\mathcal{C}_0 = \mathcal{C}^{2n+2} \circ \cdots \circ \mathcal{C}^1$ with $\mathcal{C}^j(x, y) = (x, y) + O(\mu\varepsilon^{p+j})$, $\mathcal{C}_1(x, y) = (x, y) + O(\mu^2\varepsilon^{p+3})$ and that $\mathcal{C}^1(x, y) = (x, y) + \mu\varepsilon^{p+1}(-\partial_y S_1^1, \partial_x S_1^1) + O(\mu^2\varepsilon^{2p+2})$ where $\partial_\theta S_1^1 = -h_1$ and S_1^1 has zero mean. The form of R_{2k-2} comes from (3.4), and the fact that from the second set of steps of averaging we begin with $\bar{R}_{2k-2}$ and then $R_{2k-2}^{\nu+1} = R_{2k-2}^{\nu} + \varepsilon\bar{r}_{2n-2}$. □

3.4. Estimates for the Poincaré map

3.4.1. Notation. In this section we calculate the Poincaré map associated to $\mathcal{H}$ obtained in Proposition 3.1. Let

$$\begin{aligned} x' &= y + \mu\varepsilon^{p+3}\partial_y F_{2n-2} + \mu^2\varepsilon^{p+2}\partial_y R_{2k-2} \\ y' &= -V'(x) - \mu\varepsilon^{p+3}\partial_x F_{2n-2} - \mu^2\varepsilon^{p+2}\partial_x R_{2k-2}. \end{aligned} \tag{3.11}$$

be the corresponding equation. To simplify the notation we introduce $z = (x, y)$ and

$$F_{2n-3} = (\partial_y F_{2n-2}, -\partial_x F_{2n-2}), \qquad R_{2k-3} = (\partial_y R_{2k-2}, -\partial_x R_{2k-2}).$$

We define

$$\begin{aligned} X_0(x, y) &= \begin{pmatrix} y \\ -V'(x) \end{pmatrix}, \\ Y_{\mu,\varepsilon} &= X_0 + \mu^2\varepsilon^{p+2} R_{2k-3} \end{aligned}$$

and

$$X_{\mu,\varepsilon} = Y_{\mu,\varepsilon} + \mu\varepsilon^{p+3} F_{2n-3}.$$

Hence, equation (3.11) becomes $z' = X_{\mu,\varepsilon}(z, t/\varepsilon)$.

Let $\varphi_{\mu,\varepsilon}(t, t_0, z)$ be the solution of the equation $z' = X_{\mu,\varepsilon}(z, t/\varepsilon)$ such that $\varphi_{\mu,\varepsilon}(t_0, t_0, z) = z$ and $\phi_{\mu,\varepsilon}(t, t_0, z)$ be the solution of the system $z' = Y_{\mu,\varepsilon}(z)$ such that $\phi_{\mu,\varepsilon}(t_0, t_0, z) = z$. If there is not danger of confusion, we will denote $\varphi_{\mu,\varepsilon}(t, t_0, z)$ by $\varphi_{\mu,\varepsilon}(t)$ and $\phi_{\mu,\varepsilon}(t, t_0, z)$ by $\phi_\mu(t, \varepsilon)$.

We consider the Poincaré maps

$$P_{\mu,\varepsilon}^{t_0}(z) = \varphi_{\mu,\varepsilon}(t_0 + 2\pi\varepsilon, t_0, z) \tag{3.12}$$

and

$$\hat{P}_{\mu,\varepsilon}(z) = \phi_{\mu,\varepsilon}(t_0 + 2\pi\varepsilon, t_0, z) = \phi_{\mu,\varepsilon}(2\pi\varepsilon, 0, z). \tag{3.13}$$

Let $U \subset \mathbb{C}^2$ be a neighborhood of the origin and let

$$V(\theta_0) = \left(\cup_{s\in[0,1]}\varphi_{\mu,\varepsilon}(\theta_0 + s2\pi, \theta_0, U)\right) \bigcup \left(\cup_{s\in[0,1]}\phi_{\mu,\varepsilon}(\theta_0 + s2\pi, \theta_0, U)\right)$$

and

$$V = \bigcup_{\theta_0 \in \mathbb{R}} V(\theta_0).$$

Therefore, since $\varphi_{\mu,\varepsilon}(\theta_0 + s2\pi, \theta_0, z)$ depends $2\pi\varepsilon$-periodically on θ_0, the set V is bounded.

PROPOSITION 3.2. *We have the following expressions for the Poincaré maps*

$$\begin{aligned} \hat{P}_{\mu,\varepsilon}(z) &= \begin{pmatrix} x + 2\pi\varepsilon y \\ y \end{pmatrix} + 2\pi\varepsilon \begin{pmatrix} 2\pi\varepsilon q_1(z,\varepsilon) \\ -V'(x) + 2\pi\varepsilon q_2(z,\varepsilon) \end{pmatrix} \\ &\quad + \mu^2\varepsilon^{p+3}T_{2k-3}(z) \end{aligned}$$

and

$$P^{t_0}_{\mu,\varepsilon}(z) = \hat{P}_{\mu,\varepsilon}(z) + \mu\varepsilon^{p+4}S_{2n-3}(z, t_0/\varepsilon).$$

where q_1, $q_2 \in P_{n-1}$ *(independent of* μ*),* $S_{2n-3} \in P_{2n-3}$ *and* $T_{2k-3} \in P_{2k-3}$.

3.4.2. Some preliminary bounds. In order to determine the properties of the Poincaré map defined in (3.12) we need a precise knowledge of the distance between a solution and its initial condition, the distance between the solutions of the unperturbed system, $\varphi_0(t)$, and the solutions of the perturbed one, $\varphi_{\mu,\varepsilon}(t)$, as well as the distance between $\phi_{\mu,\varepsilon}(t)$ and $\varphi_{\mu,\varepsilon}(t)$. This is studied in this subsection.

We make the convention that if $l < 0$ in $\|(x,y)\|^l$ we understand that it represents a constant term.

We need a simple lemma:

LEMMA 3.5. *Let* $\Omega \subset \mathbb{C}^2\times\mathbb{R}\times\mathbb{C}\times\mathbb{R}$ *be a neighborhood of* $\{(0,0)\}\times\mathbb{R}\times\{0\}\times\{0\}\}$ *and let* $f : \Omega \to \mathbb{R}$ *be a function that is continuous,* C^1 *with respect to* θ *and analytic with respect to* (x, y, μ) *such that there exists a constant* $c > 0$ *verifying*

$$\|f(x,y,\theta,\mu,\varepsilon)\| \leq c|y|^i\|(x,y)\|^l$$

for all $(x,y,\theta,\mu,\varepsilon) \in \Omega$. *Then there exists a function* $f_l \in P_l$, C^1 *with respect to* θ *such that*

$$f(x,y,\theta,\mu,\varepsilon) = y^i f_l(x,y,\theta,\mu,\varepsilon).$$

PROOF. We take $f_l(x,y,\theta,\mu,\varepsilon) = f(x,y,\theta,\mu,\varepsilon)/y^l$. Obviously we have to prove that f_l is analytic at points of the form $(x,0,\theta,\mu,\varepsilon) \in \Omega$. We consider $(x,0,\theta,\mu,\varepsilon) \in \Omega$ and y small enough so that the Taylor series of f with respect to y at $(x,0,\theta,\mu,\varepsilon)$ converges at y. Then by Taylor's theorem the result follows. □

It is clear that $X_{\mu,\varepsilon}$ is bounded in V and it is $2\pi\varepsilon$-periodic on t, thus there exists some constant M (independent on θ) such that, $\|X_{\mu,\varepsilon}(x,y,\theta)\| \leq M$ for all $(x,y) \in V$ and $\theta \in \mathbb{R}$.

Moreover $X_{\mu,\varepsilon}$ and $Y_{\mu,\varepsilon}$ are Lipschitz in V. We denote by L a common Lipschitz constant for $X_{\mu,\varepsilon}$ and $Y_{\mu,\varepsilon}$.

To simplify the arguments related to the dependence with respect to ε, first we will obtain estimates for the solutions of the scaled equations.

LEMMA 3.6. *Let* $\tilde{\varphi}_{\mu,\varepsilon}(\theta) = \tilde{\varphi}_{\mu,\varepsilon}(\theta, \theta_0, z)$ *be the solution of*

$$\dot{z} = \varepsilon X_{\mu,\varepsilon}(z, \theta), \qquad \tilde{\varphi}_{\mu,\varepsilon}(\theta_0) = z \tag{3.14}$$

and $\tilde{\phi}_{\mu,\varepsilon}(\theta) = \tilde{\phi}_{\mu,\varepsilon}(\theta, \theta_0, z)$ *be the solution of*

$$\dot{z} = \varepsilon Y_{\mu,\varepsilon}(z), \qquad \tilde{\phi}_{\mu,\varepsilon}(\theta_0) = z \tag{3.15}$$

Then, if $\theta \in [\theta_0, \theta_0 + 2\pi]$ *and* $z = (x, y) \in U$*, there exist some constants* C, C_F, μ_0 *and* ε_0 *such that for all* $|\mu| \leq \mu_0$ *and* $|\varepsilon| \leq \varepsilon_0$ *the following bounds hold:*

1) $\|\tilde{\varphi}_{\mu,\varepsilon}(\theta)\| \leq C\|z\|$ *and* $\|\tilde{\phi}_{\mu,\varepsilon}(\theta)\| \leq C\|z\|$.
2) $\|\tilde{\varphi}_{\mu,\varepsilon}(\theta) - z\| \leq \varepsilon C(|y| + \|z\|^{n-1} + \mu^2\varepsilon^{p+2}\|z\|^{2k-3})$.
3) *The solutions* $\tilde{\phi}_{\mu,\varepsilon}$ *and* $\tilde{\varphi}_{\mu,\varepsilon}$ *can be expressed as*

$$\tilde{\phi}_{\mu,\varepsilon}(\theta) = \varphi_0(\theta) + \mu^2\varepsilon^{p+3}\Phi_{\mu,\varepsilon}(\theta, \theta_0, z)$$

with

$$\|\Phi_{\mu,\varepsilon}(\theta, \theta_0, z)\| \leq C\|z\|^{2k-3},$$

and

$$\tilde{\varphi}_{\mu,\varepsilon}(\theta) = \tilde{\phi}_{\mu,\varepsilon}(\theta) + \mu\varepsilon^{p+4}\Psi_{\mu,\varepsilon}(\theta, \theta_0, z)$$

with

$$\|\Psi_{\mu,\varepsilon}(\theta, \theta_0, z)\| \leq C_F\|z\|^{2n-3}.$$

Furthermore, if $F_{2n-3} = 0$, $\Psi_{\mu,\varepsilon} = 0$. *Moreover,* $\Psi_{\mu,\varepsilon}$ *and* $\Phi_{\mu,\varepsilon}$ *are* C^0, C^1 *with respect to* θ *and* θ_0 *and analytic with respect to* μ *and the initial condition* z.
4) *The functions*

$$\begin{aligned} S_{2n-3}(z, \theta_0) &\equiv \Psi_{\mu,\varepsilon}(\theta_0 + 2\pi, \theta_0, z) \\ T_{2k-3}(z) &\equiv \Phi_{\mu,\varepsilon}(\theta_0 + 2\pi, \theta_0, z) \end{aligned}$$

satisfy that $S_{2n-3} \in P_{2n-3}$ *and* $T_{2k-3} \in P_{2k-3}$. *Moreover if* $F_{2n-3} = 0$, $S_{2n-3} \equiv 0$.

REMARK 3.4. *We observe that, since system (3.15) is autonomous,* T_{2k-3} *actually does not depend on* θ_0.

PROOF. The proof of estimates 1) and 2) follows from Gronwall's lemma. To deal with the third property, we look for the solutions of (3.15) in the form

$$\tilde{\phi}_{\mu,\varepsilon}(\theta) = \varphi_0(\theta) + \mu^2\varepsilon^{p+3}\Phi_{\mu,\varepsilon}(\theta, \theta_0, z).$$

We denote $\Phi_{\mu,\varepsilon}(\theta, \theta_0, z)$ by $\Phi_{\mu,\varepsilon}(\theta)$. From the identity

$$X_0(\tilde{\phi}_{\mu,\varepsilon}(\theta)) = X_0(\varphi_0(\theta)) + X_0(\varphi_0(\theta) + \mu^2\varepsilon^{p+3}\Phi_{\mu,\varepsilon}(\theta)) - X_0(\varphi_0(\theta)),$$

we deduce that

$$\dot{\Phi}_{\mu,\varepsilon} = \frac{1}{\mu^2\varepsilon^{p+2}}[X_0(\varphi_0 + \mu^2\varepsilon^{p+3}\Phi_{\mu,\varepsilon}) - X_0(\varphi_0)] + R_{2k-3}(\tilde{\phi}_{\mu,\varepsilon})$$

with the initial condition $\Phi_{\mu,\varepsilon}(\theta_0) = (0,0)$. We observe that since $\varphi_0 = \varphi_0(\theta, \theta_0, z)$ is analytic, the function $\Phi_{\mu,\varepsilon}$ is C^0, C^1 with respect to θ and θ_0, and analytic with respect to μ. From the differential equation for $\Phi_{\mu,\varepsilon}$

$$\begin{aligned}\Phi_{\mu,\varepsilon}(\theta) &= \frac{1}{\mu^2\varepsilon^{p+2}}\int_{\theta_0}^{\theta}[X_0(\varphi_0(s)+\mu^2\varepsilon^{p+3}\Phi_{\mu,\varepsilon}(s)) - X_0(\varphi_0(s))]\,ds \\ &\quad + \int_{\theta_0}^{\theta} R_{2k-3}(\tilde{\phi}_{\mu,\varepsilon}(s))\,ds\end{aligned}$$

and, since X_0 is Lipschitz and $\|\tilde{\phi}_{\mu,\varepsilon}(s)\| \le C\|z\|$ we have that

$$\begin{aligned}\|\Phi_{\mu,\varepsilon}(\theta)\| &\le L\varepsilon\int_{\theta_0}^{\theta}\|\Phi_{\mu,\varepsilon}(s)\|\,ds + \int_{\theta_0}^{\theta}\|R_{2k-3}(\tilde{\phi}_{\mu,\varepsilon}(s))\|\,ds \\ &\le L\varepsilon\int_{\theta_0}^{\theta}\|\Phi_{\mu,\varepsilon}(s)\|\,ds + C\|z\|^{2k-3}.\end{aligned}$$

An application of Gronwall's lemma gives the bound

$$\|\Phi_{\mu,\varepsilon}(\theta)\| \le Ce^{L\varepsilon 2\pi}\|z\|^{2k-3}, \qquad \theta \in [\theta_0, \theta_0+2\pi]. \tag{3.16}$$

It is clear that

$$T_{2k-3}(z) \equiv \Phi_{\mu,\varepsilon}(\theta_0+2\pi, \theta_0, z)$$

does not depend on θ_0 and it is analytic with respect to initial conditions. Therefore, by Lemma 3.5, $T_{2k-3} \in P_{2k-3}$.

Analogously, we look for solutions of (3.14) of the form

$$\tilde{\varphi}_{\mu,\varepsilon}(\theta) = \tilde{\phi}_{\mu,\varepsilon}(\theta) + \mu\varepsilon^{p+4}\Psi_{\mu,\varepsilon}(\theta, \theta_0, z).$$

As before we denote $\Psi_{\mu,\varepsilon}(\theta, \theta_0, z)$ by $\Psi_{\mu,\varepsilon}(\theta)$. The function $\Psi_{\mu,\varepsilon}$ satisfies the following equation:

$$\dot{\Psi}_{\mu,\varepsilon} = \frac{1}{\mu\varepsilon^{p+3}}[Y_{\mu,\varepsilon}(\tilde{\phi}_{\mu,\varepsilon}+\mu\varepsilon^{p+4}\Psi_{\mu,\varepsilon}) - Y_{\mu,\varepsilon}(\tilde{\phi}_{\mu,\varepsilon})] + F_{2n-2}(\tilde{\varphi}_{\mu,\varepsilon}).$$

Therefore, we obtain the estimate

$$\begin{aligned}\|\Psi_{\mu,\varepsilon}(\theta)\| &\le L\varepsilon\int_{\theta_0}^{\theta}\|\Psi_{\mu,\varepsilon}(\theta)\|\,ds + \int_{\theta_0}^{\theta}\|F_{2n-2}(\tilde{\varphi}_{\mu,\varepsilon}(s), s)\|\,ds \\ &\le L\varepsilon\int_{\theta_0}^{\theta}\|\Psi_{\mu,\varepsilon}(\theta)\|\,ds + \tilde{C}_F\|z\|^{2n-3}.\end{aligned}$$

As before, Gronwall's lemma gives the bound

$$\|\Psi_{\mu,\varepsilon}(\theta)\| \le C_F\|z\|^{2n-3}, \qquad \theta \in [\theta_0, \theta_0+2\pi]. \tag{3.17}$$

It is clear that $\Psi_{\mu,\varepsilon}$ is zero if F_{2n-2} is zero and that the function

$$S_{2n-3}(z, \theta_0) = \Psi_{\mu,\varepsilon}(\theta_0+2\pi, \theta_0, z)$$

is 2π-periodic in θ_0 and analytic with respect to z. Moreover by estimate (3.17) and Lemma 3.5, $S_{2n-3} \in P_{2n-3}$. □

PROOF OF PROPOSITION 3.2. We have that for all t, t_0 for which the solutions are defined

$$\varphi_{\mu,\varepsilon}(t, t_0, z) = \tilde{\varphi}_{\mu,\varepsilon}(t/\varepsilon, t_0/\varepsilon, z)$$

and the analogous relation for $\phi_{\mu,\varepsilon}$ and $\tilde{\phi}_{\mu,\varepsilon}$. Then we can write

$$P^{t_0}_{\mu,\varepsilon}(z) = \tilde{\varphi}_{\mu,\varepsilon}(t_0/\varepsilon + 2\pi, t_0/\varepsilon, z)$$

and

$$\hat{P}_{\mu,\varepsilon}(z) = \tilde{\phi}_{\mu,\varepsilon}(2\pi, 0, z).$$

By properties 3) and 4) of Lemma 3.6 it is enough to compute the Poincaré map of the unperturbed system, which is independent of t_0 since the unperturbed system is autonomous:

$$\begin{aligned} P^{t_0}_{0,\varepsilon}(z) &= \varphi_0(2\pi\varepsilon, 0, z) \\ &= z + 2\pi\varepsilon\varphi_0'(0) + 4\pi^2\varepsilon^2 \int_0^1 (1-s)\varphi_0''(s2\pi\varepsilon)\, ds \\ &= z + 2\pi\varepsilon \begin{pmatrix} y \\ -V'(x) \end{pmatrix} \\ &\quad + 4\pi^2\varepsilon^2 \begin{pmatrix} \int_0^1 (1-s)V'(\varphi_0^1(s2\pi\varepsilon))\, ds \\ -\int_0^1 (1-s)V''(\varphi_0^1(s2\pi\varepsilon))\varphi_0^2(s2\pi\varepsilon)\, ds \end{pmatrix}. \end{aligned}$$

It is clear that for $s \in [0,1]$

$$\begin{aligned} |V'(\varphi_0^1(s2\pi\varepsilon))| &\leq C\|z\|^{n-1} \\ |V''(\varphi_0^1(s2\pi\varepsilon))\varphi_0^2(s2\pi\varepsilon)| &\leq C\|z\|^{n-1}. \end{aligned}$$

Hence, by Lemma 3.5,

$$\begin{aligned} V'(\varphi_0^1(s2\pi\varepsilon)) &\in P_{n-1} \\ V''(\varphi_0^1(s2\pi\varepsilon))\varphi_0^2(s2\pi\varepsilon) &\in P_{n-1}. \end{aligned}$$

Therefore,

$$\begin{aligned} P^{t_0}_{0,\varepsilon}(z) &= \varphi_0(2\pi\varepsilon, 0, z) \\ &= z + 2\pi\varepsilon \begin{pmatrix} y \\ -V'(x) \end{pmatrix} + 4\pi^2\varepsilon^2 \begin{pmatrix} q_1(z,\varepsilon) \\ q_2(z,\varepsilon) \end{pmatrix}. \end{aligned}$$

□

3.4.3. The stable manifold of the auxiliary system $z' = Y_{\mu,\varepsilon}(z)$. In this subsection we compute the parameterization of the stable manifold of $z' = Y_{\mu,\varepsilon}(z)$ and we compare it with γ_0. For that we will take advantage of the fact that the vector field $Y_{\mu,\varepsilon}$ is Hamiltonian and autonomous.

PROPOSITION 3.3. *Consider the system $z' = Y_{\mu,\varepsilon}(z)$ with μ and ε small enough. Then*

1) *The system $z' = Y_{\mu,\varepsilon}(z)$ has local stable manifold. Moreover, given $\tau > 0$ there exists $T > 0$ such that it can be represented by $\hat{\gamma}(u) = (\hat{\alpha}(u), \hat{\beta}(u))$, $u \in D(T,\tau)$, with*

$$\hat{\alpha}(u) = \frac{c_1}{u^{2/(n-2)}}\left(1 + O\left(\frac{1}{u^\nu}\right)\right), \qquad c_1^{n-2} = \frac{-2}{a_n(n-2)^2} + O(\mu^2\varepsilon^{p+2})$$

and

$$\hat{\beta}(u) = \frac{c_2}{u^{n/(n-2)}}\left(1 + O\left(\frac{1}{u^\nu}\right)\right), \qquad c_2 = c_1\frac{-2}{(n-2)}$$

where $\nu > 0$.

2) *There exists* $M > 0$ *such that for* $u \in D(T,\tau)$

$$|\hat{\alpha}(u) - \alpha_0(u)| \le M \frac{\mu^2 \varepsilon^{p+2}}{|u|^{2/(n-2)}}, \qquad |\hat{\beta}(u) - \beta_0(u)| \le M \frac{\mu^2 \varepsilon^{p+2}}{|u|^{n/(n-2)}}.$$

PROOF. 1) Since $Y_{\mu,\varepsilon}(z) = J[Dh_0 + \mu^2\varepsilon^{p+2} DR_{2k-2}]$ where J is the usual symplectic matrix, we know that, if the stable manifold exists, it is contained in

$$h_0(z) + \mu\varepsilon^{p+2} R_{2k-2}(z,\mu,\varepsilon) = 0. \tag{3.18}$$

We know that when $\mu = 0$ the local stable manifold is given by $y = f_0(z) = -\sqrt{-2V(x)}$, which is defined in $\Omega(\pi, r)$. For $\mu \neq 0$ we look for the local stable manifold in the form

$$y = f_0(x) + \mu^2\varepsilon^{p+2} g(x,\mu,\varepsilon).$$

Putting this expression into (3.18) we get the following equation for g

$$\begin{aligned} & g(x,\mu,\varepsilon) \\ & = \frac{2R_{2k-2}(x, f_0(x) + \mu^2\varepsilon^{p+2} g(x,\mu,\varepsilon))}{\sqrt{-2V(x)} + \sqrt{-2V(x) - \mu^2\varepsilon^{p+2} 2R_{2k-2}(x, f_0(x) + \mu^2\varepsilon^{p+2} g(x,\mu,\varepsilon))}}. \end{aligned} \tag{3.19}$$

To solve this equation, we introduce the domain $\Omega_1 = \Omega_1(\pi, r, \mu_0, \varepsilon_0)$ and the space

$$\Sigma = \left\{ g : \Omega_1 \to \mathbb{C} : \text{ continuous, analytic in } (x,\mu), \ \sup_{\Omega_1} \frac{|g(x,\mu,\varepsilon)|}{|x|^{2k-2-n/2}} < +\infty \right\}$$

with the norm $\|g\|_\Sigma = \sup_{\Omega_1} |g(x,\mu,\varepsilon)|/|x|^{2k-2-n/2}$.

Let Γ be the operator such that Γg is defined by the right hand side of (3.19). It is easy to see that there exists a closed ball Σ_M, of radius M, in Σ such that Γ has a fixed point in Σ_M, which provides us with the solution we were looking for.

Restricting the differential equation to the local manifold we find the equation

$$x' = f_0(x) + \mu^2\varepsilon^{p+2} f_1(x,\mu,\varepsilon) \tag{3.20}$$

where

$$f_1(x) = g(x,\mu,\varepsilon) + \partial_y R_{2k-2}(x, f_0(x) + \mu^2\varepsilon^{p+2} g(x,\mu,\varepsilon),\mu,\varepsilon).$$

Now, applying Proposition 2.1 (Chapter 2) we get that a solution of (3.20) is given by

$$\hat{\alpha}(u) = \frac{c_1}{u^{2/(n-2)}}\left(1 + O\left(\frac{1}{u^\nu}\right)\right), \qquad u \in D(T,\tau)$$

with $\nu > 0$ and c_1 depending on μ, ε and satisfying

$$c_1^{n-2} = \frac{-2}{a_n(n-2)^2} + O(\mu^2\varepsilon^{p+2}).$$

$\hat{\beta}(u)$ is obtained as $\hat{\beta}(u) = f_0(\tilde{\alpha}(u)) + \mu^2\varepsilon^{p+2} g(\tilde{\alpha}(u),\mu,\varepsilon)$ and hence

$$\hat{\beta}(u) = \frac{c_2}{u^{n/(n-2)}}\left(1 + O\left(\frac{1}{u^\nu}\right)\right),$$

where $\nu > 0$ and

$$c_2 = c_1 \frac{-2}{(n-2)}.$$

All solutions have the form $\hat{\gamma}(u-u_0)$ and have the same asymptotic expressions. We choose u_0 such that $\hat{\alpha}(T)=\alpha_0(T)$.

2) We define $\xi(u)=\hat{\alpha}(u)-\alpha_0(t)$. Since $\dot{\alpha}_0(u)=f_0(\alpha_0(u))$ we have that

$$\begin{aligned}\dot{\xi} &= f_0(\hat{\alpha})-f_0(\alpha_0)+\mu^2\varepsilon^{p+2}f_1(\hat{\alpha},\mu,\varepsilon)\\ &= Df_0(\alpha_0)\xi+[f_0(\hat{\alpha})-f_0(\alpha_0)-Df_0(\alpha_0)\xi]+\mu^2\varepsilon^{p+2}f_1(\hat{\alpha},\mu,\varepsilon).\end{aligned}$$

Since $\xi(u)=\dot{\alpha}_0(u)=f_0(\alpha_0(u))$ is a solution of $\dot{\xi}=Df_0(\alpha_0)\xi$, applying the formula of variation of parameters and using that $\xi(T)=0$ we can write,

$$\begin{aligned}\xi(u) &= f_0(\alpha_0(u))\left[\int_T^u \frac{1}{f_0(\alpha_0(s))}[f_0(\hat{\alpha}(s))-f_0(\alpha_0(s))-Df_0(\alpha_0(s))\xi(s)]\,ds\right.\\ &\left.+\mu^2\varepsilon^{p+2}\int_T^u \frac{1}{f_0(\alpha_0(s))}f_1(\hat{\alpha}(s),\mu,\varepsilon)\,ds\right].\end{aligned}\tag{3.21}$$

We have the a priori estimate $|\xi(u)|\le|\hat{\alpha}(u)|+|\alpha_0(u)|\le C_1/|u|^{2/(n-2)}$ for $u\in D(T,\tau)$. Then we also have

$$|D^2f_0(\alpha_0(s)+\zeta\xi(s))|\le\frac{C_2}{|s|^{(n-4)/(n-2)}}\qquad\text{and}\qquad|f_1(\hat{\alpha}(s),\mu,\varepsilon)|\le\frac{C_3}{|s|^q}$$

with $q=\min\left\{\frac{2(2k-2-n/2)}{n-2},\frac{2(2n-3)}{n-2}\right\}\ge n/(n-2)>1$, for $s\in D(T,\tau)$. Moreover

$$\frac{\tilde{C}_4}{|s|^{n/(n-2)}}\le f_0(\alpha_0(s))\le\frac{C_4}{|s|^{n/(n-2)}}.$$

Let $t_1=\sup\left\{t\in[T,+\infty):|\xi(u)|\le\mu^2\varepsilon^{p+2}M\frac{1}{|u|^{2/(n-2)}},\operatorname{Re}u\in[T,t]\right\}$, with M to be chosen below. Assume that $t_1<+\infty$. For $t\in[T,t_1]$, using Taylor's theorem we have that

$$\begin{aligned}|f_0(\hat{\alpha}(s))-f_0(\alpha_0(s))&-Df_0(\alpha_0(s))\xi(s)|\\ &=\left|\int_0^1(1-\zeta)D^2f_0(\alpha_0(s)+\zeta\xi(s))\xi^2(s)\,d\zeta\right|\\ &\le\frac{C_5}{|s|^{(n-4)/(n-2)}}\mu^4\varepsilon^{2p+4}M^2\frac{1}{|s|^{4/(n-2)}}.\end{aligned}$$

Putting the previous bounds in (3.21) we obtain

$$\begin{aligned}|\xi(u)| &\le \frac{C_4}{|u|^{n/(n-2)}}\left[\int_T^u\frac{1}{\tilde{C}_4}|s|^{n/(n-2)}C_5\mu^4\varepsilon^{2p+4}M^2\frac{1}{|s|^{n/(n-2)}}\,ds\right.\\ &\left.+\mu^2\varepsilon^{p+2}\int_T^u\frac{1}{\tilde{C}_4}|s|^{n/(n-2)}C_3\frac{1}{|s|^q}\,ds\right]\\ &\le\mu^2\varepsilon^{p+2}[\mu^2\varepsilon^{p+2}C_6M^2+C_7]\frac{1}{|u|^{2/(n-2)}}\end{aligned}$$

with C_7 independent of M.

We choose $M=2C_7$, then, since $M>C_7$ and μ, ε are small we get that for u such that $\operatorname{Re}u=t_1$, $|\xi(u)|<\mu^2\varepsilon^{p+2}M/|u|^{2/(n-2)}$ which is a contradiction with the definition of t_1. Therefore, for $u\in\Omega(T,\tau)$

$$|\xi(u)|\le\mu^2\varepsilon^{p+2}M\frac{1}{|u|^{2/(n-2)}}.$$

The estimate for $\hat{\beta} - \beta_0$ follows from

$$\hat{\beta} - \beta_0 = f_0(\tilde{\alpha}) + \mu^2 \varepsilon^{p+2} g(\hat{\alpha}, \mu, \varepsilon) - f_0(\alpha_0).$$

□

3.5. The operators B and $\mathcal{B}$

The Banach spaces we use in this section were introduced at the beginning of Section 3.2.

We will need the operator $B_k : \mathcal{X}_k^{\mathrm{s}} \longrightarrow \mathcal{X}_k^{\mathrm{s}}$ defined by

$$(B_k \sigma)(t, s) = \sigma(t + 2\pi\varepsilon, s) - \sigma(t, s) \tag{3.22}$$

with $\varepsilon > 0$. It is a well defined linear operator with $\|B_k\| \leq 2$. Indeed, it is readily seen that if $\sigma \in \mathcal{X}_k^{\mathrm{s}}$ then $B_k\sigma \in \mathcal{X}_k^{\mathrm{s}}$ and that

$$\begin{aligned}
(t + \operatorname{Re} s)^k |(B_k\sigma)(t, s)| &\leq (t + \operatorname{Re} s)^k |\sigma(t + 2\pi\varepsilon, s)| + (t + \operatorname{Re} s)^k |\sigma(t, s)| \\
&\leq (t + 2\pi\varepsilon + \operatorname{Re} s)^k |\sigma(t + 2\pi\varepsilon, s)| \left(\frac{t + \operatorname{Re} s}{t + 2\pi\varepsilon + \operatorname{Re} s} \right)^k \\
&\quad + (t + \operatorname{Re} s)^k |\sigma(t, s)| \\
&\leq 2\|\sigma\|_k.
\end{aligned}$$

Of course, we have used that $t + \operatorname{Re} s \geq T > 0$.

REMARK 3.5. *In fact* $\|B_k\| = 2$. *For the function* $\sigma \in \mathcal{X}_k^{\mathrm{s}}$ *defined by*

$$\sigma(t, s) = \frac{1}{\cosh(a/(2\varepsilon))} \frac{1}{(t + s)^k} \sin \frac{t + s}{2\varepsilon}$$

we have $\|\sigma\|_k = 1$ *and* $\|B_k\sigma\|_k = 2$.

We are interested in finding a right inverse of the operator B_k. For that, we write $B_k\sigma = \psi$ from which we can obtain

$$\sigma(t, s) = -\psi(t, s) + \sigma(t + 2\pi\varepsilon, s). \tag{3.23}$$

Applying (3.23) iteratively

$$\sigma(t, s) = -\sum_{j=0}^{N} \psi(t + 2\pi\varepsilon j, s) + \sigma(t + 2\pi\varepsilon(N + 1), s). \tag{3.24}$$

If $\sigma \in \mathcal{X}_k^{\mathrm{s}}$, $\lim_{t\to\infty} \sigma(t, s) = 0$ so that we are allowed to take limit as $N \to \infty$ in (3.24) and we obtain the formal expression

$$(B_k^{-1}\psi)(t, s) = -\sum_{j=0}^{\infty} \psi(t + 2\pi\varepsilon j, s). \tag{3.25}$$

LEMMA 3.7. *The operator* $B_k : \mathcal{X}_k^{\mathrm{s}} \longrightarrow \mathcal{X}_k^{\mathrm{s}}$ *has right inverses* $B_k^{-1} : \mathcal{X}_m^{\mathrm{s}} \longrightarrow \mathcal{X}_k^{\mathrm{s}}$ *with* $m \geq k + 1$ *and*

$$\|B_k^{-1}\psi\|_k \leq \frac{1}{T^{m-k-1}} \left(\frac{1}{2T} + \frac{1}{2\pi\varepsilon(m - 1)} \right) \|\psi\|_m.$$

In particular, if $T \geq (m - 1)\pi/4$,

$$\|B_k^{-1}\psi\|_k \leq \frac{1 + 4\varepsilon}{2\pi\varepsilon k} \|\psi\|_m.$$

PROOF. We define $\psi_N(t,s) = \sum_{j=0}^{N} \psi(t+2\pi\varepsilon j, s)$ and

$$(B_k^{-1}\psi)(t,s) = -\lim_{N\to\infty} \psi_N(t,s).$$

Let $m \geq k+1$. First we check that if $\psi \in \mathcal{X}_m^{\mathrm{s}}$, ψ_N converges uniformly. Indeed, from

$$|\psi(t+2\pi\varepsilon j, s)| \leq \frac{1}{(t+2\pi\varepsilon j + \operatorname{Re} s)^m}\|\psi\|_m \leq \left(\frac{1}{T+2\pi\varepsilon j}\right)^m \|\psi\|_m,$$

the claim follows form the M-test of Weierstrass.

One immediately shows that $B_k^{-1}\psi$ satisfies the first three conditions which define $\mathcal{X}_k^{\mathrm{s}}$.

Moreover given $\psi \in \mathcal{X}_m^{\mathrm{s}}$

$$\begin{aligned}\|B_k^{-1}\psi\|_k &= \sup_{(t,s)\in D^{\mathrm{s}}} \sum_{j=0}^{\infty} (t+\operatorname{Re} s)^k |\psi(t+2\pi\varepsilon j, s)| \\ &\leq \sup_{(t,s)\in D^{\mathrm{s}}} \sum_{j=0}^{\infty} \frac{(t+\operatorname{Re} s)^k}{(t+2\pi\varepsilon j+\operatorname{Re} s)^m}\|\psi\|_m.\end{aligned}$$

To bound the sum we introduce $u = t + \operatorname{Re} s$ and we bound

$$\sum_{j=0}^{\infty} \frac{(t+\operatorname{Re} s)^k}{(t+2\pi\varepsilon j+\operatorname{Re} s)^m} = \frac{1}{2\pi\varepsilon}\frac{2\pi\varepsilon}{u^{m-k}} \sum_{j=0}^{\infty} \frac{1}{\left(1+\frac{2\pi\varepsilon j}{u}\right)^m}.$$

Then the sum can be bounded by

$$\begin{aligned}\frac{1}{2\pi\varepsilon}\frac{1}{u^{m-k-1}}\left[\frac{2\pi\varepsilon}{2u} + \int_0^\infty \frac{1}{(1+x)^m}\,dx\right] &= \frac{1}{2\pi\varepsilon}\frac{1}{u^{m-k-1}}\left[\frac{2\pi\varepsilon}{2u} + \frac{1}{m-1}\right] \\ &= \frac{1}{2u^{m-k}} + \frac{1}{2\pi\varepsilon(m-1)u^{m-k-1}}.\end{aligned}$$

From the definitions of both operators we easily see that

$$B_k \circ B_k^{-1} = \mathrm{Id}_{|\mathcal{X}_m^{\mathrm{s}}}.$$

□

We define $\mathcal{B} : \mathcal{X}_k^{\mathrm{s}} \times \mathcal{X}_{k+1}^{\mathrm{s}} \longrightarrow \mathcal{X}_k^{\mathrm{s}} \times \mathcal{X}_{k+1}^{\mathrm{s}}$ by

$$\mathcal{B}(\sigma_1, \sigma_2) = (B_k\sigma_1, B_{k+1}\sigma_2)$$

where B_k is defined in (3.22) and $\mathcal{B}^{-1} : \mathcal{X}_{k+1}^{\mathrm{s}} \times \mathcal{X}_{k+2}^{\mathrm{s}} \longrightarrow \mathcal{X}_k^{\mathrm{s}} \times \mathcal{X}_{k+1}^{\mathrm{s}}$ by

$$\mathcal{B}^{-1}(\psi_1, \psi_2) = (B_{k+1}^{-1}\psi_1, B_{k+2}^{-1}\psi_2)$$

where B_j^{-1} is defined in (3.25). Clearly

$$\mathcal{B}\mathcal{B}^{-1} = \mathrm{Id}_{|\mathcal{X}_{k+1}^{\mathrm{s}} \times \mathcal{X}_{k+2}^{\mathrm{s}}}.$$

3.6. Proof of Theorem 3.1

We look for the parameterization of the stable manifold of equation (3.11) in the form

$$\tilde{\gamma}^{s}_{\mu,\varepsilon}(t,s) = \hat{\gamma}(t+s) + \mu\varepsilon^{p+2}\sigma(t,s)$$

with $\sigma = (\sigma_1, \sigma_2)$ belonging to a suitable space of functions decreasing to zero at some prescribed rate. (See below the choice of the spaces.)

We introduce the notation

$$Q_{2k-3}(z) = 4\pi^2 \begin{pmatrix} q_1(z,\varepsilon) \\ q_2(z,\varepsilon) \end{pmatrix} + \mu^2\varepsilon^{p+1}T_{2k-3}(z).$$

Therefore we can write

$$\hat{P}_{\mu,\varepsilon}(z) = z + 2\pi\varepsilon \begin{pmatrix} y \\ -V'(x) \end{pmatrix} + \varepsilon^2 Q_{2k-3}(z)$$

and

$$P^{t}_{\mu,\varepsilon}(z) = \hat{P}_{\mu,\varepsilon}(z) + \mu\varepsilon^{p+4}S_{2n-3}(z, t/\varepsilon).$$

We impose the conditions

$$P^{t}_{\mu,\varepsilon}(\tilde{\gamma}^{s}_{\mu,\varepsilon}(t,s)) = \tilde{\gamma}^{s}_{\mu,\varepsilon}(t+2\pi\varepsilon, s) \tag{3.26}$$

and

$$\tilde{\gamma}^{s}_{\mu,\varepsilon}(t+2\pi\varepsilon, s) = \tilde{\gamma}^{s}_{\mu,\varepsilon}(t, s+2\pi\varepsilon).$$

Expanding by Taylor around $\hat{\gamma}$, the condition (3.26) becomes

$$\begin{aligned} &P^{t}_{\mu,\varepsilon}(\tilde{\gamma}^{s}_{\mu,\varepsilon}(t,s)) \\ &\quad = \hat{P}_{\mu,\varepsilon}(\hat{\gamma}(t+s)) + \mu\varepsilon^{p+4}S_{2n-3}(\hat{\gamma}(t+s), t/\varepsilon) + \mu\varepsilon^{p+2}D\hat{P}_{\mu,\varepsilon}(\hat{\gamma}(t+s))\sigma(t,s) \\ &\quad\quad + \mu^2\varepsilon^{2p+6}DS_{2n-3}(\hat{\gamma}(t+s), t/\varepsilon)\sigma(t,s) + \mu^2\varepsilon^{2p+4}O(|\sigma(t,s)|^2). \end{aligned}$$

Thus, (3.26) is equivalent to

$$\begin{aligned} \sigma(t+2\pi\varepsilon, s) &= \sigma(t,s) + 2\pi\varepsilon \begin{pmatrix} \sigma_2(t,s) \\ -V''(\hat{\alpha}(t+s))\sigma_1(t,s) \end{pmatrix} \\ &\quad + \varepsilon^2 DQ_{2k-3}(\hat{\gamma}(t+s))\sigma(t,s) + \varepsilon^2 S_{2n-3}(\hat{\gamma}(t+s), t/\varepsilon) \\ &\quad + \mu\varepsilon^{p+2}O(|\sigma(t,s)|^2) + \mu\varepsilon^{p+4}DS_{2n-3}(\hat{\gamma}(t+s), t/\varepsilon)\sigma(t,s). \end{aligned}$$

To simplify the notation, we further introduce

$$A(\hat{\alpha}(t+s)) = \begin{pmatrix} 0 & 1 \\ -V''(\hat{\alpha}(t+s)) & 0 \end{pmatrix},$$

$$\begin{aligned} H(\sigma)(t,s) &= DQ_{2k-3}(\hat{\gamma}(t+s))\sigma(t,s) \\ &\quad + S_{2n-3}(\hat{\gamma}(t+s), t/\varepsilon) + \mu\varepsilon^{p}O(|\sigma(t,s)|^2) \\ &\quad + \mu\varepsilon^{p+2}DS_{2n-3}(\hat{\gamma}(t+s), t/\varepsilon)\sigma(t,s) \end{aligned} \tag{3.27}$$

and

$$\mathcal{F}(\sigma)(t,s) = 2\pi\varepsilon A(\hat{\alpha}(t+s))\sigma(t,s) + \varepsilon^2 H(\sigma)(t,s). \tag{3.28}$$

Then the problem is reduced to find $\sigma = (\sigma_1, \sigma_2)$ such that

$$\mathcal{B}\sigma = \mathcal{F}(\sigma).$$

In the next lemma we will see that $\mathcal{F}$ sends $\mathcal{X}^{\mathrm{s}}_m \times \mathcal{X}^{\mathrm{s}}_{m+1}$ to $\mathcal{X}^{\mathrm{s}}_{m+1} \times \mathcal{X}^{\mathrm{s}}_{m+2}$ and hence we can use a right inverse of $\mathcal{B}$, as given in Section 3.5, to obtain the fixed point equation

$$\sigma = \mathcal{B}^{-1}\mathcal{F}(\sigma).$$

We look for $\sigma \in \mathcal{X}^{\mathrm{s}}_m \times \mathcal{X}^{\mathrm{s}}_{m+1}$ with a suitable value of m. We define the following weighted norm in the product space $\mathcal{X}^{\mathrm{s}}_m \times \mathcal{X}^{\mathrm{s}}_{m+1}$:

$$\|(h_m, h_{m+1})\|_m = L\|h_m\|_m + \|h_{m+1}\|_{m+1}$$

with

$$L = \frac{n-1}{n-2} + \frac{n^2}{(3n-4)(n-2)}$$

and we denote $B(r, m, m+1) \subset \mathcal{X}^{\mathrm{s}}_m \times \mathcal{X}^{\mathrm{s}}_{m+1}$ the closed ball of radius r with this norm.

LEMMA 3.8. *Let $m = \frac{2n-2}{n-2}$. There exists $r > 0$ independent of μ and ε, such that the operator*

$$\mathcal{B}^{-1} \circ \mathcal{F} : B(r, m, m+1) \longrightarrow B(r, m, m+1)$$

is well defined and is a contraction.

PROOF. We recall that, by Proposition 3.3, the function $\hat{\gamma} \in \mathcal{X}^{\mathrm{s}}_{2/(n-2)} \times \mathcal{X}^{\mathrm{s}}_{n/(n-2)}$ and hence, if $f_l \in P_l$,

$$f_l \circ \hat{\gamma} \in \mathcal{X}^{\mathrm{s}}_{2l/(n-2)}.$$

Let $\sigma \in B(r, m, m+1)$. Since $V'' \circ \hat{\alpha} \in \mathcal{X}^{\mathrm{s}}_2$, it is clear that

$$A(\hat{\alpha})\sigma \in \mathcal{X}^{\mathrm{s}}_{m+1} \times \mathcal{X}^{\mathrm{s}}_{m+2}. \tag{3.29}$$

We check that

$$\begin{aligned} S_{2n-3}(\hat{\gamma}(t+s), t/\varepsilon) &\in \mathcal{X}^{\mathrm{s}}_{m+2} \times \mathcal{X}^{\mathrm{s}}_{m+2} \\ (DQ_{2k-3} \circ \hat{\gamma})\,\sigma &\in \mathcal{X}^{\mathrm{s}}_{m+1} \times \mathcal{X}^{\mathrm{s}}_{m+2} \\ DS_{2n-3}(\hat{\gamma}(t+s), t/\varepsilon)\,\sigma(t,s) &\in \mathcal{X}^{\mathrm{s}}_{m+4} \times \mathcal{X}^{\mathrm{s}}_{m+4}. \end{aligned} \tag{3.30}$$

Indeed, we recall that if $S_{2n-3} \in P_{2n-3}$, then $S_{2n-3}(\hat{\gamma}(t+s), t/\varepsilon) \in \mathcal{X}^{\mathrm{s}}_{m+2} \times \mathcal{X}^{\mathrm{s}}_{m+2}$.

Now we deal with $(DQ_{2k-3} \circ \hat{\gamma})\,\sigma$. We denote

$$DQ_{2k-3} \circ \hat{\gamma} = \begin{pmatrix} \hat{Q}_{11} & \hat{Q}_{12} \\ \hat{Q}_{21} & \hat{Q}_{22} \end{pmatrix}.$$

It is clear that, since all elements of DQ_{2k-3} belong to P_{n-2}, we have that $\hat{Q}_{ij} \in \mathcal{X}^{\mathrm{s}}_{2(n-2)/(n-2)} = \mathcal{X}^{\mathrm{s}}_2$ and therefore

$$(DQ_{2k-3} \circ \hat{\gamma})\sigma \in \mathcal{X}^{\mathrm{s}}_{m+2} \times \mathcal{X}^{\mathrm{s}}_{m+2}.$$

To deal with $(DS_{2n-3} \circ \hat{\gamma})\sigma$, we observe that $DS_{2n-3} \circ \hat{\gamma} \in \mathcal{X}^{\mathrm{s}}_4$ and therefore

$$DS_{2n-3}(\hat{\gamma}(t,s), t/\varepsilon)\sigma(t,s) \in \mathcal{X}^{\mathrm{s}}_{m+4} \times \mathcal{X}^{\mathrm{s}}_{m+4}.$$

Thus, by (3.30) and since $O(|\sigma|^2) \in \mathcal{X}^{\mathrm{s}}_{2m+2}$, we have that

$$H(\sigma) \in \mathcal{X}^{\mathrm{s}}_{m+1} \times \mathcal{X}^{\mathrm{s}}_{m+2}. \tag{3.31}$$

Therefore, by (3.29), (3.31) and by definition (3.28) of $\mathcal{F}$ we have that

$$\mathcal{F}(\sigma) = 2\pi\varepsilon A(\hat{\alpha})\sigma + \varepsilon^2 H(\sigma) \in \mathcal{X}^{\rm s}_{m+1} \times \mathcal{X}^{\rm s}_{m+2}.$$

Finally, by Lemma 3.7,

$$\mathcal{B}^{-1}(\mathcal{F}(\sigma)) \in \mathcal{X}^{\rm s}_{m} \times \mathcal{X}^{\rm s}_{m+1}.$$

Next we will prove that $\|\mathcal{B}^{-1}(\mathcal{F}(\sigma))\|_m \le r$ if $\|\sigma\|_m \le r$. We begin by bounding the norms $\|\mathcal{F}_1\|_{m+1}$ and $\|\mathcal{F}_2\|_{m+2}$. From (3.27), which defines H, it is clear that there exist constants M_1 and M_2 such that

$$\|H_1(\sigma)\|_{m+1} \le M_1, \qquad \|H_2(\sigma)\|_{m+2} \le M_2.$$

Then

$$\begin{aligned}\|\mathcal{F}_1(\sigma)\|_{m+1} &\le 2\pi\varepsilon \sup_{(t,s)\in D} |\sigma_2(t,s)(t+\operatorname{Re} s)^{m+1}| \\ &\quad +\varepsilon^2 \sup_{(t,s)\in D} |H_1(\sigma)(t,s)(t+\operatorname{Re} s)^{m+1}| \\ &\le 2\pi\varepsilon\|\sigma_2\|_{m+1} + \varepsilon^2 M_1\end{aligned}$$

and, using the expression of $\hat{\alpha}$ obtained in Proposition 3.3

$$\begin{aligned}\|\mathcal{F}_2(\sigma)\|_{m+2} &\le 2\pi\varepsilon \sup_{(t,s)\in D} |V''(\hat{\alpha}(t+s))\sigma_1(t,s)(t+\operatorname{Re} s)^{m+2}| \\ &\quad +\varepsilon^2 \sup_{(t,s)\in D} |H_2(\sigma)(t,s)(t+\operatorname{Re} s)^{m+2}| \\ &\le 2\pi\varepsilon\left(\frac{2n(n-1)}{(n-2)^2} + O(T^{-\nu})\right)\|\sigma_1\|_m + \varepsilon^2 M_2,\end{aligned}$$

with $\nu > 0$. Therefore, using Lemma 3.7 we obtain

$$\begin{aligned}\|\mathcal{B}^{-1}\circ\mathcal{F}(\sigma)\|_m &= \|(\mathcal{B}_1^{-1}\mathcal{F}_1(\sigma), \mathcal{B}_2^{-1}\mathcal{F}_2(\sigma))\|_m \\ &= L\|\mathcal{B}_1^{-1}\mathcal{F}_1(\sigma_1,\sigma_2)\|_m + \|\mathcal{B}_2^{-1}\mathcal{F}_2(\sigma_1,\sigma_2)\|_{m+1} \\ &\le L\|\mathcal{F}_1\|_{m+1}\frac{1+4\varepsilon}{2m\pi\varepsilon} + \|\mathcal{F}_2\|_{m+2}\frac{1+4\varepsilon}{2(m+1)\pi\varepsilon} \\ &\le L\|\sigma_2\|_{m+1}\frac{n-2}{2n-2} + \left(\frac{2n(n-1)}{(3n-4)(n-2)} + O(T^{-\nu})\right)\|\sigma_1\|_m \\ &\quad +O(\varepsilon) \\ &\le L\frac{n-2}{2n-2}\|\sigma_2\|_{m+1} + \frac{2n^2}{(3n-4)(n-2)}\|\sigma_1\|_m + O(\varepsilon)\end{aligned} \tag{3.32}$$

if T is big, in particular $O(T^{-\nu}) \le \frac{2n}{(3n-4)(n-2)}$. We introduce

$$\begin{aligned}a &= \frac{2(n-1)}{n-2} \\ b &= \frac{2n^2}{(3n-4)(n-2)}.\end{aligned}$$

We observe that since $a > b$ and $L = (a+b)/2$ we have $b < L < a$. Thus, $1 - La^{-1}$and $L - b$ are positive numbers. Moreover if we introduce the constant

$K = \frac{2a}{a-b} > 2$ we have that

$$\begin{aligned} L - b &= \frac{a-b}{2} > \frac{L}{K} \\ 1 - La^{-1} &= \frac{a-b}{2a} = \frac{1}{K}. \end{aligned}$$

Therefore, we can bound (3.32) as follows

$$\begin{aligned} La^{-1}\|\sigma_2\|_{m+1} + b\|\sigma_1\|_m + O(\varepsilon) &\leq (\|\sigma_2\|_{m+1} + L\|\sigma_1\|_m)(1 - \frac{1}{K}) + O(\varepsilon) \\ &\leq \left(1 - \frac{1}{K}\right) r + O(\varepsilon) \\ &< r \end{aligned}$$

if ε is small enough.

Now, we prove that $\mathcal{B}^{-1} \circ \mathcal{F}$ is a contraction. Let $\sigma = (\sigma_1, \sigma_2) \in B(r, m, m+1)$ and $\bar{\sigma} = (\bar{\sigma}_1, \bar{\sigma}_2) \in B(r, m, m+1)$. From (3.27) it is easy to see that

$$\|H(\sigma) - H(\bar{\sigma})\|_m \leq C\|\sigma - \bar{\sigma}\|_m$$

for some constant $C > 0$. Thus, by definition (3.28) of $\mathcal{F}$, we obtain, doing similar estimations as above

$$\begin{aligned} &\|\mathcal{B}^{-1} \circ \mathcal{F}(\sigma_1, \sigma_2) - \mathcal{B}^{-1} \circ \mathcal{F}(\bar{\sigma}_1, \bar{\sigma}_2)\|_m \\ &\quad \leq L\frac{n-2}{2n-2}\|\sigma_2 - \bar{\sigma}_2\|_{m+1} + \frac{2n^2}{(3n-4)(n-2)}\|\sigma_1 - \bar{\sigma}_1\|_m + \varepsilon C\|\sigma - \bar{\sigma}\|_m \\ &\quad \leq \left(1 - \frac{1}{2K}\right) \|\sigma - \bar{\sigma}\|_m \end{aligned}$$

if ε is small enough.

Then, applying the fixed point theorem we obtain the existence and uniqueness of a fixed point $(\sigma_1, \sigma_2) \in B(r, m, m+1)$. This ends the proof of the lemma. □

Now we can finish the proof of Theorem 3.1.

END OF THE PROOF OF THEOREM 3.1. The parameterization of the local stable manifold so far obtained needs not to be a solution of equation (3.1) with respect to t. To have a parameterization which is a solution with respect to t we take $t_0 = T - 2\pi\varepsilon$, with $T - 2\pi\varepsilon$ satisfying the previous results and we define

$$\chi^{\mathrm{s}}_{\mu,\varepsilon}(t,s) = \varphi_{\mu,\varepsilon}(t, t_0, \tilde{\gamma}^{\mathrm{s}}_{\mu,\varepsilon}(t_0, s)), \qquad t > T - 2\pi\varepsilon, \quad \mathrm{Re}\, s > -2\pi\varepsilon, \quad |\mathrm{Im}\, s| \leq \tau,$$

where here $\varphi_{\mu,\varepsilon}(t, t_0, x, y)$ is the general solution of equation (3.1).

For $t > T - 2\pi\varepsilon, \mathrm{Re}\, s > -2\pi\varepsilon$ we have

$$\begin{aligned} \chi^{\mathrm{s}}_{\mu,\varepsilon}(t, s + 2\pi\varepsilon) &= \varphi_{\mu,\varepsilon}(t, t_0, \tilde{\gamma}^{\mathrm{s}}_{\mu,\varepsilon}(t_0, s + 2\pi\varepsilon)) = \varphi_{\mu,\varepsilon}(t, t_0, \tilde{\gamma}^{\mathrm{s}}_{\mu,\varepsilon}(t_0 + 2\pi\varepsilon, s)) \\ &= \varphi_{\mu,\varepsilon}(t + 2\pi\varepsilon, t_0 + 2\pi\varepsilon, \varphi_{\mu,\varepsilon}(t_0 + 2\pi\varepsilon, t_0, \tilde{\gamma}^{\mathrm{s}}_{\mu,\varepsilon}(t_0, s))) \\ &= \varphi_{\mu,\varepsilon}(t + 2\pi\varepsilon, t_0, \tilde{\gamma}^{\mathrm{s}}_{\mu,\varepsilon}(t_0, s)) \\ &= \chi^{\mathrm{s}}_{\mu,\varepsilon}(t + 2\pi\varepsilon, s). \end{aligned}$$

This relation permits to extend $\chi^{\mathrm{s}}_{\mu,\varepsilon}$ to D^{s}. The extension is a solution of equation (3.1) with respect to t and is analytic with respect to s, and clearly satisfies 2).

Now we will check that for $(t,s) \in D^{\mathrm{s}}$, $\chi^{\mathrm{s}}_{\mu,\varepsilon}(t,s) = \gamma_0(t+s) + \mu\varepsilon^{p+2} r(t,s)$ with $r(t,s) = O(\frac{1}{|t+\operatorname{Re} s|^{2/(n-2)}})$. Indeed, let $k \in \mathbb{Z}$ such that $|t - 2\pi\varepsilon k - T| < 2\pi\varepsilon$.

$$\begin{aligned}
\chi^{\mathrm{s}}_{\mu,\varepsilon}(t, s+2\pi\varepsilon) &= \varphi_{\mu,\varepsilon}(t-2\pi\varepsilon k, T-2\pi\varepsilon k, \tilde{\gamma}^{\mathrm{s}}_{\mu,\varepsilon}(T,s)) \\
&= \varphi_{\mu,\varepsilon}(t-2\pi\varepsilon k, T-2\pi\varepsilon k, \varphi_{\mu,\varepsilon}(T-2\pi\varepsilon k, T, \tilde{\gamma}^{\mathrm{s}}_{\mu,\varepsilon}(T+2\pi\varepsilon k, s))) \\
&= \varphi_{\mu,\varepsilon}(t-2\pi\varepsilon k, T, \tilde{\gamma}^{\mathrm{s}}_{\mu,\varepsilon}(T+2\pi\varepsilon k, s)) \\
&= \phi_{\mu,\varepsilon}(t-2\pi\varepsilon k, T, \tilde{\gamma}^{\mathrm{s}}_{\mu,\varepsilon}(T+2\pi\varepsilon k, s)) + \mu\varepsilon^{p+4} O((\tilde{\gamma}^{\mathrm{s}}_{\mu,\varepsilon})^{2n-3}) \\
&= \phi_{\mu,\varepsilon}(t-2\pi\varepsilon k, T, \hat{\gamma}(T+2\pi\varepsilon k+s)) + \mu\varepsilon^{p+2} O(\sigma) + \mu\varepsilon^{p+4}((\tilde{\gamma}^{\mathrm{s}}_{\mu,\varepsilon})^{2n-3}) \\
&= \hat{\gamma}(t+s) + \mu\varepsilon^{p+2} O\Big(\frac{1}{|t+\operatorname{Re} s|^{(2n-2)/(n-2)}}\Big) \\
&= \gamma_0(t+s) + \mu\varepsilon^{p+2} O\Big(\frac{1}{|t+\operatorname{Re} s|^{2/(n-2)}}\Big).
\end{aligned}$$

Going back to the original variables we obtain the result we have stated in Theorem 3.1. Indeed, by Proposition 3.1 the change $\mathcal{C}$ has the form

$$(x,y) = \mathcal{C}(\bar{x}, \bar{y}, t/\varepsilon) = (\bar{x}, \bar{y}) + \mu\varepsilon^{p+1} G(\bar{x}, \bar{y}, t/\varepsilon) + O(\mu\varepsilon^{p+2}).$$

Finally we take

$$\begin{aligned}
\gamma^{\mathrm{s}}_{\mu,\varepsilon}(t,s) &= \mathcal{C}(\chi^{\mathrm{s}}_{\mu,\varepsilon}(t,s), t/\varepsilon) \\
&= \chi^{\mathrm{s}}_{\mu,\varepsilon}(t,s) + \mu\varepsilon^{p+1} G(\chi^{\mathrm{s}}_{\mu,\varepsilon}(t,s), t/\varepsilon) + \mu\varepsilon^{p+2} O(\chi^{\mathrm{s}}_{\mu,\varepsilon}(t,s)^{k-1}) \\
&= \gamma_0(t+s) + \mu\varepsilon^{p+1} G(\gamma_0(t+s), t/\varepsilon) + O(\mu\varepsilon^{p+2}).
\end{aligned}$$

We observe that $\gamma^{\mathrm{s}}_{\mu,\varepsilon}$ is defined for all $(t,s) \in D^{\mathrm{s}}$, and that conclusion 2) follows because the change $\mathcal{C}$ is $2\pi\varepsilon$-periodic in t. □

4. Flow box coordinates

4.1. Introduction

In this chapter we prove the existence of flow box coordinates of a quasi integrable system under general hypotheses, in a neighborhood of the stable manifold of the unperturbed system, which does not contain the origin and it is independent of the parameters μ and ε. A similar result is in [**Ge2**]. There, the flow box coordinates are found implicitly using the variational equations in a neighborhood of the stable manifold. Our proof gives these coordinates in an explicit way and gives a careful estimate of the distance between the change of coordinates in the unperturbed case and the change in the perturbed one, using variational equations with respect to the parameter μ.

In [**DS2**] [**DS1**] the authors use flow-box coordinates defined near a hyperbolic fixed point. To construct such coordinates they use the Birkhoff Normal Form in an essential way. Also, in [**DG**], the authors construct flow box coordinates near a n-dimensional hyperbolic invariant tori, but only in real domains.

We begin by introducing notation and the hypotheses **H1** and **H2** we will assume in this chapter. With these hypotheses we will prove a result on the existence of flow box coordinates: Theorem 4.1. Then, the application of this theorem to equation (1.1) of Chapter 1 gives Theorem 4.2, in which the result is obtained applying Theorem 4.1, not directly to (1.1) but to the averaged equation.

To prove Theorem 4.1 first, in Section 4.3, we translate the stable manifold to the first axis of coordinates and in these coordinates, for the unperturbed system, we construct explicitly the flow box coordinates, just integrating the equation and using that the system is Hamiltonian.

To construct the flow box coordinates for the general system in a neighborhood of (a part of) the stable manifold, we use a special parameterization of the solutions of the equation. We parameterize the solutions, $z(t, s, Y)$ with two parameters $(t, s) \in \mathbb{R} \times \mathbb{C}$ in such a way that $t \in \mathbb{R}$ is a time parameter, $Y \in \mathbb{C}$ and

$$z(t + 2\pi\varepsilon, s, Y) = z(t, s + 2\pi\varepsilon, Y)$$

in a suitable domain. To obtain this we use a technique designed by Lazutkin to do a controlled analytic continuation. This is done in Subsections 4.5.2 and 4.5.3. From this parameterization we easily obtain another parameterization of the form

$$w(t + s, t/\varepsilon, Y),$$

that is, we separate in some way the slow time $t + s$ and the fast time t/ε.

Next we find a first flow box coordinates $(\mathcal{T}, \mathcal{Y})$ from the condition

$$w(\mathcal{T}(x, v, \theta), \theta, \mathcal{Y}(x, v, \theta)) = (x, v)$$

using the scheme of the proof of the implicit function theorem. We obtain them close to the analogous ones we have calculated in the non perturbed case. Then easily we pass to new flow box coordinates $(\mathcal{T}, \mathcal{F})$ with $\mathcal{F}$ close to the energy variable (the Hamiltonian).

Finally, using the Hamiltonian character of the equations, we slightly modify these coordinates to make them canonical.

4.2. Definitions and main result

We consider Hamiltonian systems of the form

$$H(x,y,t/\varepsilon) = H_0(x,y) + \mu\varepsilon^q H_1(x,y,t/\varepsilon,\mu,\varepsilon)$$

where

$$H_0(x,y) = \frac{y^2}{2} + V(x).$$

REMARK 4.1. *Since we will apply the result of existence of flow box coordinates to an averaged system, here q and H_1 mean a generic exponent and a generic Hamiltonian respectively which (in general) do not coincide with p and h_1 introduced in Chapter 1.*

The associated equations of the Hamiltonian H are

$$\begin{aligned} \dot{x} &= y + \mu\varepsilon^q \partial_y H_1(x,y,t/\varepsilon,\mu,\varepsilon) \\ \dot{y} &= -V'(x) - \mu\varepsilon^q \partial_x H_1(x,y,t/\varepsilon,\mu,\varepsilon). \end{aligned} \tag{4.1}$$

For $w = (w_1, w_2) \in \mathbb{C}^2$, we define

$$\|w\| = \max\{|w_1|, |w_2|\}.$$

We will assume the following hypotheses

H1 The potential V is an analytic function in $\{x \in \mathbb{C} : |x| < \rho_0\}$, $V(x) = a_n x^n + \dots$ with $a_n < 0$, $n \in \mathbb{N}$ and $n \geq 3$.

H2 $H_1(x,y,\theta,\varepsilon,\mu)$ is C^0, 2π-periodic in θ and analytic in the x, y, μ variables. The variables (x,y,μ) belong to

$$\{(x,y) \in \mathbb{C}^2 : \|(x,y)\| < \rho_0\} \times \{\mu \in \mathbb{C} : |\mu| < \mu_0\},$$

$\theta \in \mathbb{R}$ and $0 < \varepsilon < \varepsilon_0$. Moreover $H_1(x,y,\theta,\varepsilon,\mu) = O(\|(x,y)\|^k)$ with $k \geq 2$.

In the applications of the results of this chapter, k will be always greater of equal than $n-1$.

For the unperturbed system, the origin has a stable manifold which can be represented as the graph of the function

$$f(x) = -\sqrt{-2V(x)}.$$

We observe that f is analytic in

$$\Omega(\rho_0, \pi) = \{x \in \mathbb{C} : |x| < \rho_0\} \setminus \{\operatorname{Re} x \leq 0,\ \operatorname{Im} x = 0\}.$$

REMARK 4.2. *By Proposition 2.2 (Chapter 2), the unperturbed system ($\mu = 0$) has a parameterization of the stable manifold defined in*

$$\{u \in \mathbb{C} : |u| \geq T,\ \ |\arg u| < \pi\}$$

which we denote by $\gamma_0(u) = (\alpha_0(u), \beta_0(u))$.

In the next definition we fix some parameters.

DEFINITION 4.1. *Let C_0 and C_1 be such that*

$$C_0|x|^\beta \leq |f(x)| \leq C_1|x|^\beta$$

for all $x \in \Omega(\rho_0, \pi)$ where $\beta = n/2$.

Let $\kappa_0 \in \mathbb{R}$ be as in Proposition 2.5. We take $\delta_0 = \alpha_0(\kappa_0)$. We recall that to take κ_0 bigger implies to take δ_0 smaller. Given $\delta_0' > 0$ such that $\delta_0' < \delta_0$, we define

$$r_0 = \frac{1}{2}\Big(- C_1\delta_0^\beta + \sqrt{C_1^2\delta_0^{2\beta} + C_0^2(\delta_0')^{2\beta}}\Big).$$

In the following we take δ_0' (and consequently r_0) small enough.

For $\kappa_1^- > \kappa_0$, $\kappa_1^+ > \kappa_1^-$, $\kappa_2 > 0$, $\kappa_3 > 0$ such that $\kappa_1^- - \kappa_3 > \kappa_0$ and $0 < r < r_0$, we define the open sets

$$\begin{aligned} D_0(\kappa_1^\pm, \kappa_2, \kappa_3) &= \{s \in \mathbb{C} : \kappa_1^- - \kappa_3 < \operatorname{Re} s < \kappa_1^+ + \kappa_3, \; |\operatorname{Im} s| < \kappa_2 + \kappa_3\} \\ D^*(\kappa_1^\pm, \kappa_2, \kappa_3) &= \alpha_0(D_0(\kappa_1^\pm, \kappa_2, \kappa_3)) \\ V_0(\kappa_1^\pm, \kappa_2, \kappa_3, r) &= \{(x, y) \in \mathbb{C}^2 : x \in D^*(\kappa_1^\pm, \kappa_2, \kappa_3), \; |y - f(x)| < r\} \end{aligned}$$

and

$$V(\kappa_1^\pm, \kappa_2, \kappa_3, r) = \{(x, y, \theta) \in \mathbb{C}^2 \times \mathbb{R} : (x, y) \in V_0(\kappa_1^\pm, \kappa_2, \kappa_3, r)\}.$$

In the following we will consider some changes of variables depending on (x, y, θ), and we will refer to them as canonical changes, meaning that are canonical as a function of (x, y) with θ fixed.

The main result of this Chapter is:

THEOREM 4.1. *(Flow box coordinates). Let $\kappa_1^- > 2\kappa_0$. If hypotheses **H1**, **H2** hold, for any $\kappa_1^+ > \kappa_1^-$ and $\kappa_2 > 0$ there exists $r > 0$ and a canonical change of variables*

$$(x, y, \theta = t/\varepsilon) \in V(\kappa_1^\pm, \kappa_2, 0, r) \mapsto (T, I, \theta) = (\mathcal{T}^1(x, y, \theta), \mathcal{I}^1(x, y, \theta), \theta) \in \mathcal{V}$$

of class C^1, 2π-periodic in θ and analytic in the x, y variables, such that it transforms system (4.1) to

$$\begin{aligned} \dot{T} &= 1 \\ \dot{I} &= 0 \end{aligned}$$

and satisfies

$$\mathcal{T}^1(x, y, \theta) = \mathcal{T}_0(x, y) + O(\mu\varepsilon^q), \qquad \mathcal{I}^1(x, y, \theta) = \mathcal{I}_0(x, y) + O(\mu\varepsilon^q),$$

where the change $(x, y) \mapsto (\mathcal{T}_0(x, y), \mathcal{I}_0(x, y))$ is the corresponding change for the unperturbed system. Moreover the change is continuous in $(x, y, \theta, \mu, \varepsilon)$ and analytic in (x, y, μ).

To study the splitting of separatrices we will use the next theorem which is a specialized version of Theorem 4.1 when we apply it to a system of the form

$$\begin{aligned} x' &= y + \mu\varepsilon^{p+3}\partial_y F_{2n-2} + \mu^2\varepsilon^{p+2}\partial_y R_{2k-2} \\ y' &= -V'(x) - \mu\varepsilon^{p+3}\partial_x F_{2n-2} - \mu^2\varepsilon^{p+2}\partial_x R_{2k-2} \end{aligned}$$

which comes from system (1.1) of Chapter 1 by the averaging procedure described in Chapter 3. If system (1.1) satisfies hypotheses **HP1-HP4**, the results of Chapter 3 apply and we have that the stable manifold exists and can be parameterized by $\gamma^{\mathrm{s}}_{\mu,\varepsilon}(t, s)$.

Next theorem gives a new flow box coordinates and additional information over the values of these flow box coordinates on the stable manifold $\gamma^{\mathrm{s}}_{\mu,\varepsilon}(t, s)$. Let $\mathcal{C}$ be the change which transforms system (1.1) of Chapter 1 to the averaged system which is given in Proposition 3.1.

THEOREM 4.2. *Given $\kappa_1^- > 2\kappa_0$, $\kappa_1^+ > \kappa_1^-$ and $\kappa_2 > a$, there exist $r > 0$ and a canonical change of variables*

$$(x, y, \theta = t/\varepsilon) \in \mathcal{C}(V(\kappa_1^\pm, \kappa_2, 0, r)) \mapsto (S, E, \theta) = (\mathcal{S}(x, y, \theta), \mathcal{E}(x, y, \theta), \theta) \in \mathcal{V}_1$$

of class C^1, 2π-periodic in θ and analytic in the x, y variables, such that it transforms system (1.1) of Chapter 1 to

$$\dot{S} = 1$$
$$\dot{E} = 0$$

and satisfies

$$\mathcal{S}(x, y, \theta) = \mathcal{S}_0(x, y) + O(\mu\varepsilon^{p+1}), \qquad \mathcal{E}(x, y, \theta) = \mathcal{E}_0(x, y) + O(\mu\varepsilon^{p+1})$$

where $(x, y) \mapsto (\mathcal{S}_0(x, y), \mathcal{E}_0(x, y))$ is the corresponding change when $\mu = 0$. In fact

$$\mathcal{E}_0(x, y) = h_0(x, y).$$

Moreover, given $T \geq 0$ big enough there exist $\kappa_1^- > 2\kappa_0$, $\kappa_1^+ > \kappa_1^-$ and $\kappa_2 > a$ such that for all (t, s) such that $T \leq |t + \operatorname{Re} s| \leq 2T$ and $|\operatorname{Im} s| < a$, the parameterization $\gamma^{\mathrm{s}}_{\mu,\varepsilon}(t, s)$ of the local stable manifold given in Chapter 3 satisfies

$$\gamma^{\mathrm{s}}_{\mu,\varepsilon}(t, s) \in \mathcal{C}(V(\kappa_1^\pm, \kappa_2, 0, r))$$

and, for any $t_0 \in \mathbb{R}$

$$\mathcal{S}(\gamma^{\mathrm{s}}_{\mu,\varepsilon}(t, s), t/\varepsilon) = t - t_0 + s + \mu\varepsilon^{p+1}\mathcal{X}(s) \qquad \text{and} \qquad \mathcal{E}(\gamma^{\mathrm{s}}_{\mu,\varepsilon}(t, s), t/\varepsilon) = 0$$

and $\mathcal{X}(s_0) = 0$ for some s_0, which we can choose freely, depending on initial conditions on the stable curve. Moreover $\mathcal{X}(s)$ is analytic and $2\pi\varepsilon$-periodic.

In addition the change $(x, y, \theta) \mapsto (S, E, \theta)$ is continuous in $(x, y, \theta, \mu, \varepsilon)$ and analytic in (x, y, μ).

REMARK 4.3. *To fix ideas we consider only the parabolic case, which is the object of this memoir, but the proof also would be adapted to the hyperbolic case, with some small changes.*

In this chapter, at some places, we omit the dependence on μ and ε which is assumed that it is analytical and continuous respectively.

4.3. A preliminary change of variables

Since the stable manifold of the unperturbed system can be expressed as the graph of an analytic function, we can easily move it to the x-axis.

For any $\kappa_1^- > 2\kappa_0$, $\kappa_1^+ > \kappa_1^-$, $\kappa_2 > 0$ and $\kappa_3 > 0$ such that $\kappa_1^- - 3\kappa_3 > \kappa_0$ and $r \leq r_0$, we define the sets

$$U_0(\kappa_1^\pm, \kappa_2, \kappa_3, r) = \left\{(x, v) \in \mathbb{C}^2 : x \in D^*(\kappa_1^\pm, \kappa_2, \kappa_3, r),\ |v| < r\right\}$$

and

$$U(\kappa_1^\pm, \kappa_2, \kappa_3, r) = \left\{(x, v, \theta) \in \mathbb{C}^2 \times \mathbb{R} : (x, v) \in U_0(\kappa_1^\pm, \kappa_2, \kappa_3, r)\right\}.$$

We perform the change of variables $C : V(\kappa_1^\pm, \kappa_2, 3\kappa_3, r_0) \to \mathbb{C}^2 \times \mathbb{R}$ defined by

$$(x, y, \theta) \mapsto (x, v = y - f(x), \theta). \tag{4.2}$$

This change is canonical. It maps $V(\kappa_1^\pm, \kappa_2, 3\kappa_3, r_0)$ onto $U(\kappa_1^\pm, \kappa_2, 3\kappa_3, r_0)$.

We note that, in general, we can not extend C in such a way that it is analytic at $x = 0$ because f is not analytic at 0. The equations in these new variables are

$$
\begin{aligned}
\dot{x} &= v + f(x) + \mu\varepsilon^q \tilde{g}_1(x, v, \theta) \\
\dot{v} &= -vf'(x) + \mu\varepsilon^q \tilde{g}_2(x, v, \theta) \\
\dot{\theta} &= 1/\varepsilon
\end{aligned}
\tag{4.3}
$$

where

$$
\begin{aligned}
\tilde{g}_1(x, v, \theta) &= \partial_y H_1(x, v + f(x), \theta) \\
\tilde{g}_2(x, v, \theta) &= -\partial_x H_1(x, v + f(x), \theta) - f'(x)\partial_y H_1(x, v + f(x), \theta).
\end{aligned}
$$

We denote by $X : U(\kappa_1^{\pm}, \kappa_2, 3\kappa_3, r_0) \to \mathbb{C}^2 \times \mathbb{R}$, the vector field $X = X_0 + \mu\varepsilon^q X_1$ with

$$
X_0 = \begin{pmatrix} v + f(x) \\ -vf'(x) \\ 1/\varepsilon \end{pmatrix} \quad \text{and} \quad X_1 = \begin{pmatrix} \tilde{g}_1(x, v, \theta) \\ \tilde{g}_2(x, v, \theta) \\ 0 \end{pmatrix}.
$$

Along this section, K will denote a generic constant independent of μ, ε.

Some preliminary bounds of the vector field X are necessary.

LEMMA 4.1. *We have that*

1) *For all* $(x, v) \in U_0(\kappa_1^{\pm}, \kappa_2, 3\kappa_3, r_0)$ *and* $h \in \mathbb{R}^2$ *so small that the segment* $\overline{(x, v), (x, v) + h} \in U_0(\kappa_1^{\pm}, \kappa_2, 3\kappa_3, r_0)$,

$$
\|X_1(x + h_1, v + h_2) - X_1(x, v)\| \leq K\delta_0^{k-2}\|h\|
$$

and

$$
\|X_0(x + h_1, v + h_2) - X_0(x, v) - DX_0(x, v)h\| \leq K\|h\|^2
$$

where K *depends on the domain* $U_0(\kappa_1^{\pm}, \kappa_2, 3\kappa_3, r_0)$.

2) *There exists a constant* K *such that for all* $(x, v, \theta) \in \overline{U(\kappa_1^{\pm}, \kappa_2, 3\kappa_3, r_0)}$

$$
\|X_1(x, v, \theta)\| \leq K\|(x, v)\|^{k-1}.
$$

PROOF. The proof is straightforward. It is only necessary to use that if $x \in \overline{U(\kappa_1^{\pm}, \kappa_2, 3\kappa_3, r_0)}$, then $|x| \leq 2\delta_0$, that $f(x) = -\sqrt{-2V(x)} = O(x^{n/2})$, that $H_1(x, y, \theta)$ is a function of order k and finally, we use that the origin does not belong to the domain $\overline{U(\kappa_1^{\pm}, \kappa_2, 3\kappa_3, r_0)}$. The condition on the segment, permits to apply the mean value theorem. □

4.4. The unperturbed case

When $\mu = 0$, system (4.3) is Hamiltonian with Hamiltonian

$$
\mathcal{F}_0(x, v) = \frac{v^2}{2} + vf(x). \tag{4.4}
$$

Then for any initial condition $z^0 = (x^0, v^0, \theta^0)$ the corresponding solution is contained in the curve

$$
v = -f(x) \pm \sqrt{f^2(x) + 2\mathcal{F}_0(x^0, v^0)}
$$

when one has to choose the sign in such a way that the relation is satisfied by the initial condition. From system (4.3) it is clear that

$$
\dot{x} = \pm\sqrt{(v^0)^2 + 2v^0 f(x^0) - 2V(x)}. \tag{4.5}
$$

Let $\psi_0(t, x, v)$ be the flow of the unperturbed Hamiltonian system and let δ be such that

$$\alpha_0(\kappa_1^+) < \delta < \alpha_0(\kappa_1^-).$$

From now on δ will be fixed. Integrating equation (4.5) we find the time (in general complex time) to arrive from (x^*, v^*) to (x, v) where $x^* = \delta$ and v^* is determined by the energy conservation. In this way we get that the functions $\mathcal{T}_0(x, v)$, $\mathcal{Y}_0(x, v)$ defined in $U_0(\kappa_1^\pm, \kappa_2, 3\kappa_3, r_0)$ by

$$\mathcal{T}_0(x, v) = -\int_x^\delta \frac{ds}{\sqrt{2\mathcal{F}_0(x, v) - 2V(s)}} \tag{4.6}$$

$$\mathcal{Y}_0(x, v) = -f(\delta) + \sqrt{f^2(\delta) + 2\mathcal{F}_0(x, v)} \tag{4.7}$$

are such that $\psi_0(\mathcal{T}_0(x, v), \delta, \mathcal{Y}_0(x, v)) = (x, v)$. We choose the sign minus in (4.5), because it is obvious that, in the real case and over the stable manifold, $x(t)$ must decrease as t goes to $+\infty$. In the coordinates

$$(T, Y) = (\mathcal{T}_0(x, v), \mathcal{Y}_0(x, v))$$

the equations of the unperturbed system become:

$$\begin{aligned} \dot{T} &= 1 \\ \dot{Y} &= 0. \end{aligned}$$

Also we can consider the change

$$(x, v) \in U_0(\kappa_1^\pm, \kappa_2, 3\kappa_3, r_0) \mapsto (\mathcal{T}_0(x, v), \mathcal{F}_0(x, v)) \in \mathcal{V},$$

where $\mathcal{F}_0$ is the Hamiltonian. The equations in the coordinates

$$(T, F) = (\mathcal{T}_0(x, v), \mathcal{F}_0(x, v))$$

also are

$$\begin{aligned} \dot{T} &= 1 \\ \dot{F} &= 0. \end{aligned}$$

This second change is canonical, i.e. $\partial_x\mathcal{T}_0\partial_v\mathcal{F}_0 - \partial_v\mathcal{T}_0\partial_x\mathcal{F}_0 = 1$.

4.5. Flow box coordinates in a complex domain

4.5.1. Introduction and definitions. Let $z_0(u, Y)$ be the solution of the unperturbed Hamiltonian system $\mathcal{F}_0$, given in (4.4) such that $z_0(0, Y) = (\delta, Y)$.

When $Y = 0$, $z_0(u, 0)$ parameterizes a piece of the stable manifold near $(x, v) = (\delta, 0)$. Therefore, writing $z_0 = (x_0, v_0)$, there exists $\tau \in \mathbb{R}$ such that

$$x_0(u, 0) = \alpha_0(u + \tau) = \frac{c}{(u+\tau)^p} + O(|u + \tau|^{-\nu}), \qquad p = 2/(n-2), \nu > 0. \tag{4.8}$$

Note that $\delta = \alpha_0(\tau)$. The condition $\delta < \alpha_0(\tau_1^-)$ means that $\tau > \kappa_1^-$.

For any $\tau_1^\pm$, τ_2, τ_3 and $r_1 < r_0$, we define the sets

$$\begin{aligned} D(\tau_1^\pm, \tau_2, \tau_3) &= \{(t, s) \in \mathbb{R} \times \mathbb{C} : s, t + s \in D_0(\tau_1^\pm, \tau_2, \tau_3) \text{ and } |t| < 4\pi\varepsilon\} \\ W(r_1) &= \{Y \in \mathbb{C} : |Y| < r_1\}. \end{aligned}$$

We observe that, if $(t, s) \in D(\tau_1^\pm, \tau_2, \tau_3)$ the segment $\overline{(0, s), (t, s)}$ is contained in $D(\tau_1^\pm, \tau_2, \tau_3)$.

LEMMA 4.2. *Let* $\kappa_1^- > 2\kappa_0$ *and* $\kappa_2 > 0$. *Then, given* $\kappa_1^+ > \kappa_1^-$ *and* $\kappa_3 > 0$ *such that* $\kappa_1^- - 3\kappa_3 > \kappa_0$, *there exists* $r \le r_1 < r_0/2$ *such that the image of* $\Delta = D_0(\kappa_1^\pm - \tau, \kappa_2, \kappa_3) \times W(r_1)$ *by* z_0 *contains*

$$U_0(\kappa_1^\pm, \kappa_2, 0, r)$$

and is contained in

$$U_0(\kappa_1^\pm, \kappa_2, 2\kappa_3, r_0/2).$$

PROOF. By (4.8), the image of $D_0(\kappa_1^\pm - \tau, \kappa_2, 0) \times \{0\}$ by z_0 is $U_0(\kappa_1^\pm, \kappa_2, 0, 0)$. By continuity of z_0 it follows that if $\tilde{r}_1 > 0$ is small enough, the x component of the image of Δ by z_0 contains $D^*(\kappa_1^\pm, \kappa_2, 0)$ and is contained in $D^*(\kappa_1^\pm, \kappa_2, 2\kappa_3)$.

On the other hand the solutions $z_0(u, Y) = (x(u, Y), v(u, Y))$ stay in the energy level $\mathcal{F}_0(z(0, Y)) = Y^2/2 + Yf(\delta)$. Therefore,

$$v(u, Y) = -f(x(u, Y)) + \sqrt{Y^2 + 2Yf(\delta) + f^2(x(u, Y))}$$

and hence, while $x(u, Y)$ belongs to $D^*(\kappa_1^\pm, \kappa_2, 2\kappa_3)$,

$$K_1|Y| \le |v(u, Y)| \le K_2|Y|$$

and thus there exists $r, r_1 \le \tilde{r}_1$ satisfying the stated properties. □

Our goal is to find flow box coordinates in $U(\kappa_1^\pm, \kappa_2, 0, r)$. We will find the solutions of equations (4.3) parameterized in the form

$$(z(t, s, Y), t/\varepsilon)$$

with

$$z(t, s, Y) = z_0(t + s, Y) + \mu\varepsilon^q z_1(t, s, Y),$$

$z(0, 0, Y) = (\delta, Y)$ and the additional property

$$z(t + 2\pi\varepsilon, s, Y) = z(t, s + 2\pi\varepsilon, Y).$$

This relation permits to give a dynamical interpretation of the parameter s: the iterations of the Poincaré map simply consists in increasing the variable s by $2\pi\varepsilon$. To get the solutions in this form we will rewrite (4.3) in the form

$$\begin{aligned} \dot{z} &= A(t + s)z + b(z)(t, s) \\ \dot{\theta} &= 1/\varepsilon \end{aligned}$$

and we will apply the fixed point theorem to a suitable operator in a Banach space. To construct this operator we will need another operator which we call increment operator. This operator was introduced by Lazutkin in [**La2**].

Next, we will prove that, as in the unperturbed case, the solutions with initial condition in $U(\kappa_1^\pm, \kappa_2, 0, r)$ arrive at $x = \delta$. Then we will prove that the flow can be straightened in $U(\kappa_1^\pm, \kappa_2, 0, r)$.

Finally, we will construct another change in order to get that the composition of changes is canonical.

4.5.2. Increment operator and analytic continuation. Let h, $\tau_1^\pm$, τ_2 and $D_0 = D_0(\tau_1^\pm, \tau_2, 0)$ and $W = W(r_1)$ as in the previous subsection. We define D in a slightly more general way:

$$\begin{aligned} D &= D(h, \tau_1^\pm, \tau_2) \\ &= \{(t,s) \in \mathbb{R} \times \mathbb{C} : |t| < 2h,\ \tau_1^- < \operatorname{Re} s, t + \operatorname{Re} s < \tau_1^+,\ |\operatorname{Im} s| < \tau_2\}. \end{aligned}$$

We consider the equation

$$\dot z = A(t+s)z + b(t,s,Y) \tag{4.9}$$

(here $\cdot$ denotes derivative with respect to t), where $A(u)$ is a 2×2 matrix whose elements are analytic in $D_0 = D_0(\tau_1^\pm, \tau_2, 0)$ and continuous in $\overline{D_0}$. The function $b : \overline{D} \times W \to \mathbb{C}^2$ is continuous and, for any t such that $|t| \leq 2h$, $b(t, ., .)$ is analytic. We assume that b verifies

$$b(t+h, s, Y) = b(t, s+h, Y) \tag{4.10}$$

and we look for solutions $z(t,s,Y)$ of (4.9) analytic with respect to s and Y, and satisfying

$$z(t+h, s, Y) = z(t, s+h, Y).$$

Let $M(u)$ be a fundamental matrix of the homogeneous equation

$$\frac{d}{du}\zeta = A(u)\zeta.$$

By the general theory of linear equations, M is analytic in D_0 and there exists a constant C_M such that

$$|M(u)| \leq C_M, \qquad |M^{-1}(u)| \leq C_M, \qquad u \in D_0. \tag{4.11}$$

By the variation of parameters method, the solutions of (4.9) can be expressed as

$$z(t,s,Y) = M(t+s)\Big[M^{-1}(s)c(s,Y) + \int_0^t M^{-1}(\xi+s)b(\xi,s,Y)\,d\xi\Big] \tag{4.12}$$

where $c(s,Y)$ is an arbitrary function. Therefore, if the function $c(s,Y)$ is analytic in $D_0 \times W$, $z(t,s,Y)$ given in (4.12) is continuous in $\overline{D} \times W$ and analytic with respect to (s,Y).

We write

$$z(t+h,s,Y) = M(t+h+s)\Big[M^{-1}(s)c(s,Y) + \int_0^{t+h} M^{-1}(\xi+s)b(\xi,s,Y)\,d\xi\Big].$$

Also

$$\begin{aligned} z(t,s+h,Y) &= M(t+s+h)\Big[M^{-1}(s+h)c(s+h,Y) \\ &\quad + \int_0^t M^{-1}(\xi+s+h)b(\xi,s+h,Y)\,d\xi\Big] \\ &= M(t+s+h)\Big[M^{-1}(s+h)c(s+h,Y) \\ &\quad + \int_h^{t+h} M^{-1}(\xi+s)b(\xi,s,Y)\,d\xi\Big] \end{aligned}$$

where we have made the obvious change of variables in the integral, and we have used (4.10).

We introduce the auxiliary function $f(s,Y) = M^{-1}(s)c(s,Y)$. We have that

$$z(t+h,s,Y) = z(t,s+h,Y)$$

if and only if

$$f(s,Y) - f(s+h,Y) = -\int_0^h M^{-1}(\xi+s)b(\xi,s,Y)\,d\xi.$$

Therefore, it is natural to study the operator

$$\triangle_h f(s,Y) = f(s+h,Y) - f(s,Y).$$

We want to find analytic solutions of the equation

$$\triangle_h f(s,Y) = g(s,Y), \qquad s, s+h \in D_0,\ Y \in W \tag{4.13}$$

where g is analytic in $D_0 \times W$ and continuous in $\overline{D_0} \times W$.

We define the auxiliary open sets

$$D_0^- = \{s \in \mathbb{C} : \operatorname{Re} s < \tau_1^+,\ |\operatorname{Im} s| < \tau_2\}$$

and

$$D_0^+ = \{s \in \mathbb{C} : \tau_1^- < \operatorname{Re} s,\ \ |\operatorname{Im} s| < \tau_2\}.$$

It is clear that $D_0 = D_0^+ \cap D_0^-$. For any open set $\Omega \subset \mathbb{C}$, we define the function space

$$\mathcal{A}(\Omega,W) = \{H : \overline{\Omega} \times W \to \mathbb{C} : H \text{ is analytic in } \Omega \times W \text{ and continuous in } \overline{\Omega} \times W\}.$$

The main idea of what was developed by Lazutkin in [**La2**] is the following. Construct two analytic functions $g^+ \in \mathcal{A}(D_0^+, W)$ and $g^- \in \mathcal{A}(D_0^-, W)$ such that

$$g = g^+ + g^- \quad \text{in } D_0 \times W. \tag{4.14}$$

Then, because of the linearity of equation (4.13), the problem of finding the function f can be reduced to two simpler problems: to find two functions f^+ and f^- of $\mathcal{A}(D_0^+, W)$ and $\mathcal{A}(D_0^-, W)$ respectively such that

$$\triangle_h f^\pm = g^\pm.$$

Therefore, since the operator $\triangle_h$ is linear, the function

$$f = f^+ + f^-,$$

which is defined in $(D_0^+ \times W) \cap (D_0^- \times W) = D_0 \times W$, satisfies the equation:

$$\triangle_h f(s,Y) = \triangle_h f^+(s,Y) + \triangle_h f^-(s,Y) = g^+(s,Y) + g^-(s,Y) = g(s,Y)$$

if $s, s+h \in D_0$, $Y \in W$.

To follow the previous program the first thing we must do is to construct functions $g^\pm$, defined in the corresponding extended domain and verifying (4.14). This is done by using the next lemma which also provides useful bounds of the norm of $g^\pm$ in terms of the norm of g.

LEMMA 4.3. *Let $\chi : \mathbb{C} \longrightarrow \mathbb{C}$ be a Lipschitz bounded function such that*

$$\operatorname{supp}\chi = \{\xi \in \mathbb{C} : \operatorname{Re}\xi \le \sigma\}.$$

Let

$$\Omega = \{\xi \in \mathbb{C} :\ s_1 < \operatorname{Re}\xi < s_2,\ |\operatorname{Im}\xi| < \tau_2\},$$

with $s_1 \in \mathbb{R}$, $s_1 < \sigma$ *and* $s_2 \in [\sigma, \infty) \cup \{\infty\}$. *Let* $\Omega^* = \Omega \cap \overset{\circ}{\overline{\operatorname{supp}\chi}}$ *(small circle denotes topological interior) and let* $g \in \mathcal{A}(\Omega^*, W)$. *We define*

$$h(\xi,\eta) = \frac{1}{2\pi i}\int_{\partial\Omega\cap \operatorname{supp}\chi} \frac{\chi(\zeta)}{\zeta-\xi} g(\zeta,\eta)\, d\zeta = \frac{1}{2\pi i}\int_{\partial\Omega^*} \frac{\chi(\zeta)}{\zeta-\xi} g(\zeta,\eta)\, d\zeta.$$

Then

1) h *is analytic on* $\Omega \times W$,
2) h *extends continuously to* $\overline{\Omega} \times W$,
3) *if* $(\xi_0,\eta_0) \in (\partial\Omega \cap \operatorname{supp}\chi) \times W$

$$\lim_{(\xi_0,\eta_0)} h(\xi,\eta) = \chi(\xi_0) g(\xi_0,\eta_0) + \frac{1}{2\pi i}\int_{\partial\Omega^*} \frac{\chi(\zeta)-\chi(\xi_0)}{\zeta-\xi_0} g(\zeta,\eta_0)\, d\zeta,$$

and if $(\xi_0,\eta_0) \in (\partial\Omega \cap (\operatorname{supp}\chi)^c) \times W$

$$\lim_{(\xi_0,\eta_0)} h(\xi,\eta) = \frac{1}{2\pi i}\int_{\partial\Omega^*} \frac{\chi(\zeta)-\chi(\xi_0)}{\zeta-\xi_0} g(\zeta,\eta)\, d\zeta,$$

4) *if* $(\xi,\eta) \in \Omega \times K$, *where* K *is a compact subset of* W *we have*

$$|h(\xi,\eta)| \le \left(\|\chi\| + \frac{1}{2\pi} \operatorname{Lip}\chi \ \operatorname{length}(\partial\Omega^*) \right) \|g\|_K$$

where $\|\chi\| = \sup\{|\chi(\xi)| : \xi \in \mathbb{C}\}$ *and* $\|g\|_K = \sup\{|g(\xi,\eta)| : (\xi,\eta) \in \partial\Omega^* \times K\}$.

The same results hold in the case $\operatorname{supp}\chi = \{\xi \in \mathbb{C};\ \operatorname{Re}\xi \ge \sigma\}$, $s_1 \in \{-\infty\} \cup (-\infty, \sigma]$, $s_2 \in \mathbb{R}$, $s_2 > \sigma$.

REMARK 4.4. *We observe that, in order to apply this result, we only need that the function* g *to be analytic in a bounded complex rectangle.*

This Lemma is a parameter (with respect to η) version of a lemma by Lazutkin in [**La2**]. The proof of the present version of the lemma can be found in [**Fo3**]. Using the previous technical lemma, we will construct a right inverse of the operator $\triangle_h$.

LEMMA 4.4. *Let* $D_0 = D_0(\tau_1^{\pm}, \tau_2, 0)$. *Then there is a continuous operator*

$$\triangle_h^{-1} : \mathcal{A}(D_0, W) \to \mathcal{A}(D_0, W)$$

such that given $g \in \mathcal{A}(D_0, W)$, $f = \triangle_h^{-1} g$ *is a solution of the equation*

$$f(s+h, Y) - f(s, Y) = g(s, Y) \qquad \text{for } s, s+h \in D_0,\ Y \in W \tag{4.15}$$

and its operator norm verifies $\|\triangle_h^{-1}\| \le C_{D_0} e^{h/\tau_2} h^{-1}$, *where the constant* C_{D_0} *only depends on the size of the domain* D_0.

PROOF. Let $\chi : \mathbb{R} \to [0,1]$, be the Lipschitz function defined by

$$\chi(u) = \begin{cases} 1 & \text{if } u \le \tau_1^- \\ 1 - \frac{u-\tau_1^-}{\tau_1^+ - \tau_1^-} & \text{if } \tau_1^- < u < \tau_1^+ \\ 0 & \text{if } u \ge \tau_1^+. \end{cases}$$

Let $\chi_+(s) = \chi(\operatorname{Re} s)$ and $\chi_-(s) = 1 - \chi_+(s)$, defined in $\mathbb{C}$. We observe that

$$\operatorname{supp}\chi_+ = \{s \in \mathbb{C} : \operatorname{Re} s \le \tau_1^+\}$$

and

$$\operatorname{supp}\chi_- = \{s \in \mathbb{C} : \operatorname{Re} s \ge \tau_1^-\}.$$

Moreover it is clear that $D_0 = D_0^+ \cap \overset{\circ}{\overline{\operatorname{supp}\chi_+}}$ and $D_0 = D_0^- \cap \overset{\circ}{\overline{\operatorname{supp}\chi_-}}$. Let $\rho = \tau_2^{-1}$ and $g \in \mathcal{A}(D_0, W)$. By Lemma 4.3, the functions

$$g_\pm(s,Y) = \frac{1}{2\pi i}\frac{1}{\cosh(\rho s)}\int_{\partial D_0}\frac{\chi_\pm(\xi)\cosh(\rho\xi)}{\xi - s}g(\xi,Y)\,d\xi$$

belong to $\mathcal{A}(D_0^\pm, W)$ respectively. Moreover, by 4) of Lemma 4.3, we have

$$\begin{aligned} |g_\pm(s,Y)| &\le [\|\chi_\pm\| + \frac{1}{2\pi}\operatorname{Lip}\chi_\pm \operatorname{length}(\partial D_0)]\frac{\|g\|}{|\cosh(\rho s)|}\max_{\xi\in\partial D_0}|\cosh(\rho\xi)| \\ &\le C_{D_0}\|g\|\frac{1}{|\cosh(\rho s)|} \qquad \text{for } (s,Y) \in D_0^\pm \times W \end{aligned} \tag{4.16}$$

where $\|g\|$ means $\sup_{D_0\times W}|g(s,Y)|$.

Now we construct the inverse of $\triangle_h$. Given $(s,Y) \in D_0^+ \times W$, we define

$$f_+(s,Y) = -\sum_{k\ge 0} g_+(s+kh, Y).$$

A direct substitution shows that f_+ satisfies (4.15) in $D_0^+ \times W$. In the same way, if $(s,Y) \in D_0^- \times W$,

$$f_-(s,Y) = \sum_{k\ge 1} g_-(s-kh, Y)$$

satisfies (4.15) in $D_0^- \times W$.

Until the end of the proof, C_{D_0} will be a generic constant which may take different values in different formulas but only depends on D_0 and $\tau_1 = \max\{|\tau_1^+|, |\tau_1^-|\}$.

These series are convergent. Indeed, from (4.16) we have, for $(s,Y) \in D_0^+ \times W$

$$\begin{aligned} |f_+(s,Y)| \le \sum_{k\ge 0}|g_+(s+kh,Y)| &\le C_{D_0}\|g\|\sum_{k\ge 0}\frac{1}{|\cosh(\rho(s+kh))|} \\ &\le C_{D_0}\|g\|\sum_{k\ge 0}\frac{1}{\cosh(\rho(\operatorname{Re}s + kh))|\cos(\rho\operatorname{Im}s)|} \\ &\le C_{D_0}\|g\|\frac{2}{\cos(1)}e^{-\rho\operatorname{Re}s}\sum_{k\ge 0}e^{-\rho hk} \\ &\le C_{D_0}\|g\|\frac{1}{1-e^{-\rho h}}e^{-\rho\operatorname{Re}s} \\ &\le C_{D_0}\|g\|h^{-1}e^{\rho h}e^{-\rho\operatorname{Re}s}. \end{aligned}$$

In the same way we obtain, for $(s,Y) \in D_0^- \times W$

$$|f_-(s,Y)| \le C_{D_0}\|g\|h^{-1}e^{\rho h}e^{\rho\operatorname{Re}s}.$$

Now we consider the function $f : D_0 \times W \to \mathbb{C}$, defined by $f = f_+ + f_-$. It is clear that

$$\triangle_h f(s,Y) = \triangle_h f_+(s,Y) + \triangle_h f_-(s,Y) = g_+(s,Y) + g_-(s,Y) = g(s,Y)$$

for $s, s+h \in D_0$, $Y \in W$. Moreover, since $\rho = \tau_2^{-1}$, on $D_0 \times W$,

$$|f(s,Y)| \le C_{D_0}\|g\|h^{-1}e^{(\tau_1+h)/\tau_2} = C_{D_0}\|g\|h^{-1}e^{h/\tau_2}.$$

Then the f so constructed solves (4.15). □

4.5.3. A useful parameterization of the solutions of system (4.3). In this subsection we give a good parameterization of a set of solutions passing through $x = \delta$, of the system associated to the vector field X. We introduce an additional parameter, $s \in \mathbb{C}$, to be able to reach $\{x = \delta\}$ and to obtain useful properties of the parameterization.

We recall that we denote by $z_0(u, Y)$ the solution of the unperturbed system

$$\begin{aligned} \dot{x} &= v + f(x) \\ \dot{v} &= -vf'(x) \end{aligned}$$

such that $z_0(0, Y) = (\delta, Y)$. In a similar way as in Lemma 4.2 we can prove that the image by z_0 of $\Delta_1 = D_0(\kappa_1^{\pm} - \tau, \kappa_2, 3\kappa_3) \times W(2r_1)$ is contained in $U_0(\kappa_1^{\pm}, \kappa_2, 3\kappa_3, r_0)$ (we recall that $r_1 < r_0/2$). Moreover, z_0 is analytic on Δ_1 and continuous on its boundary. Therefore, since the vector field X_0 is analytic on $U_0(\kappa_1^{\pm}, \kappa_2, 3\kappa_3, r_0)$, a fundamental matrix $M(u)$ of the system

$$\frac{d}{du} z = DX_0(z_0(u, Y))z$$

is analytic on $D_0(\kappa_1^{\pm} - \tau, \kappa_2, 2\kappa_3)$ and continuous on its boundary. Moreover as we pointed out in (4.11), $M(u)$ and $M^{-1}(u)$ are bounded in $D_0(\kappa_1^{\pm} - \tau, \kappa_2, 2\kappa_3)$.

Now we present the parameterization of the solutions of system (4.3).

PROPOSITION 4.1. *If ε and μ are small enough then a set of solutions of equation (4.3) can be expressed as parameterized curves*

$$(z(t, s, Y), t/\varepsilon) = (x(t, s, Y), v(t, s, Y), t/\varepsilon)$$

with $(t, s, Y) \in \tilde{U}$ defined by

$$\tilde{U} = D(\kappa_1^{\pm} - \tau, \kappa_2, 2\kappa_3) \times W(2r_1),$$

satisfying the following properties:

1) *$t \mapsto z(t, s, Y)$ is a solution of system (4.3).*
2) *$z(t, s, Y)$ is C^1 and analytic in (s, Y).*
3) *$z(t + 2\pi\varepsilon, s, Y) = z(t, s + 2\pi\varepsilon, Y)$.*
4) *The solution of system (4.3) is of the form*

$$z(t, s, Y) = z_0(t + s, Y) + \mu\varepsilon^q z_1(t, s, Y)$$

with

$$\sup_{\tilde{U}} |z_1(t, s, Y)| \leq K.$$

5) *For all $Y \in W(2r_1)$, $z(0, 0, Y) = (\delta, Y)$.*

PROOF. If

$$(z(t, s, Y), t/\varepsilon) = (z_0(t + s, Y) + \mu\varepsilon^q z_1(t, s, Y), t/\varepsilon)$$

is a solution of equation (4.3), where z_0 is a solution of the unperturbed equation, it is clear that

$$\dot{z}_1 = DX_0(z_0(t + s, Y))z_1 + b(z_1)(t, s, Y) \tag{4.17}$$

with

$$\begin{aligned} b(z_1)(t,s,Y) &= \frac{1}{\mu\varepsilon^q}\big[X_0(z(t,s,Y)) - X_0(z_0(t+s,Y)) \\ &\quad -\mu\varepsilon^q DX_0(z_0(t+s,Y))z_1(t,s,Y)\big] \\ &\quad +X_1(z(t,s,Y),t/\varepsilon). \end{aligned} \tag{4.18}$$

Thus, z_1 is a solution of (4.17) if and only if

$$z_1(t,s,Y) = M(t+s)\left[M^{-1}(s)c(s,Y) + \int_0^t M^{-1}(\sigma+s)b(z_1)(\sigma,s,Y)\,d\sigma\right] \tag{4.19}$$

where $M(u)$ is a fundamental matrix of the homogeneous system. At this point $c(s,Y)$ is an arbitrary function. We choose the function $c(s,Y)$ as follows. We consider

$$g(z_1)(s,Y) = -\int_0^{2\pi\varepsilon} M^{-1}(\sigma+s)b(z_1)(\sigma,s,Y)\,d\sigma \tag{4.20}$$

and we take

$$c(z_1)(s,Y) \equiv c(s,Y) = M(s)\triangle_{2\pi\varepsilon}^{-1}g(z_1)(s,Y) \tag{4.21}$$

where $\triangle_{2\pi\varepsilon}^{-1}$ is the operator defined in Lemma 4.4. This choice of $c(s,Y)$ is the one which will permit us to check that an operator to be defined below is well defined in its domain.

We define Σ to be the space of functions $z_1 : \tilde{U} \to \mathbb{C}^2$ such that $z_1 \in \Sigma$ if and only if z_1 satisfies

(a) $z_1(t,s,Y)$ is C^0 and analytic on (s,Y).
(b) For all $(t,s,Y) \in \tilde{U}$ such that $(t+2\pi\varepsilon, s) \in D(\kappa_1^\pm - \tau, \kappa_2, 2\kappa_3)$, we have that

$$z_1(t+2\pi\varepsilon,s,Y) = z_1(t,s+2\pi\varepsilon,Y).$$

(c) $\|z_1\| = \sup_{\tilde{U}} |z_1(t,s,Y)| < +\infty$.

We endow Σ with the supremum norm and it becomes a Banach space. For any $\rho > 0$, we define $\Sigma(\rho)$ as the closed ball of radius ρ of Σ. We define the operator $\mathcal{G} : \Sigma(\rho) \to \Sigma(\rho)$ to be the right hand side of (4.19):

$$\mathcal{G}(z_1)(t,s,Y) = M(t+s)\left[M^{-1}(s)c(z_1)(s,Y) + \int_0^t M^{-1}(\sigma+s)b(z_1)(\sigma,s,Y)\,d\sigma\right]$$

with $c(z_1)$ chosen as (4.21). Our goal is to prove that if ρ is suitably chosen then $\mathcal{G}$ has a fixed point in $\Sigma(\rho)$. For that we will see that $\mathcal{G}$ is well defined and that it is a contraction in $\Sigma(\rho)$.

First, we prove that $\mathcal{G}$ is well defined. Let $z_1 \in \Sigma(\rho)$. If ρ is small,

$$z(t,s,Y) \in U(\kappa_1^\pm, \kappa_2, 3\kappa_3, r_0)$$

and thus, the function $b(z_1)$ given in (4.18) is well defined. Moreover, it is clear that, since $M(t+s)$, $M^{-1}(t+s)$ and $z_1(t,s,Y)$ are C^0 and analytic in (s,Y), the function g defined in (4.20) is analytic on $\tilde{U}$. Therefore, by Lemma 4.4, the function $c(z_1)(s,Y)$ is analytic in $\tilde{U}$. Thus $\mathcal{G}(z_1)(t,s,Y)$ is also C^0 and analytic in (s,Y).

Now we prove that the property (b) holds for $\mathcal{G}(z_1)$. It is clear that, since $z_1 \in \Sigma$ and $X_1(x, v, \theta)$ is 2π-periodic in θ,

$$b(z_1)(t + 2\pi\varepsilon, s, Y) = b(z_1)(t, s + 2\pi\varepsilon, Y).$$

Then

$$\begin{aligned} \mathcal{G}(z_1)(t + 2\pi\varepsilon, s, Y) &= M(t + 2\pi\varepsilon + s)\Big[M^{-1}(s)c(s, Y) \\ &\quad + \int_0^{t+2\pi\varepsilon} M^{-1}(\sigma + s)b(z_1)(\sigma, s, Y)\, d\sigma\Big] \\ &= M(t + s + 2\pi\varepsilon)\Big[M^{-1}(s)c(s, Y) \\ &\quad + \int_{-2\pi\varepsilon}^{t} M^{-1}(\sigma + 2\pi\varepsilon + s)b(z_1)(\sigma, s + 2\pi\varepsilon, Y)\, d\sigma\Big] \end{aligned}$$

and

$$\begin{aligned} \mathcal{G}(z_1)(t, s + 2\pi\varepsilon, Y) &= M(t + s + 2\pi\varepsilon)\Big[M^{-1}(s + 2\pi\varepsilon)c(s + 2\pi\varepsilon, Y) \\ &\quad + \int_0^t M^{-1}(\sigma + 2\pi\varepsilon + s)b(z_1)(\sigma, s + 2\pi\varepsilon, Y)\, d\sigma\Big]. \end{aligned}$$

Thus, $\mathcal{G}(z_1)(t, s + 2\pi\varepsilon, Y) = \mathcal{G}(z_1)(t + 2\pi\varepsilon, s, Y)$ if and only if

$$\begin{aligned} &M^{-1}(s)c(s, Y) - M^{-1}(s + 2\pi\varepsilon)c(s + 2\pi\varepsilon, Y) \\ &\qquad = -\int_0^{2\pi\varepsilon} M^{-1}(\sigma + s)b(z_1)(\sigma, s, Y)\, d\sigma. \end{aligned}$$

This last equality holds by definition of c in (4.21).

Next we will see that if we choose ρ in a suitable way, $\mathcal{G}(\Sigma(\rho)) \subset \Sigma(\rho)$. Indeed, let C_M be a constant such that $\|M(u)\|$, $\|M^{-1}(u)\| \le C_M$. We recall that,

$$f(s, Y) = M^{-1}(s)c(s, Y).$$

By Lemmas 4.1 and 4.4 we obtain

$$\|f\| \le C_{D_0}\frac{e^{2\pi\varepsilon/\tau_2}}{2\pi\varepsilon}\|g(z_1)\| \le C_M C_{D_0}\|b(z_1)\| \le C_M C_{D_0}[K|\mu|\varepsilon^q\|z_1\|^2 + K\delta_0^{k-1}].$$

Thus

$$\begin{aligned} \|\mathcal{G}(z_1)\| &\le C_M^2 C_{D_0}(K|\mu|\varepsilon^q\rho^2 + K\delta_0^{k-1}) + C_M^2 4\pi\varepsilon(K|\mu|\varepsilon^q\rho^2 + K\delta_0^{k-1}) \\ &\le \rho \end{aligned}$$

if $\rho = 2C_M^2 K(C_{D_0} + 4\pi\varepsilon)\delta_0^{k-1}$ and $|\mu|\varepsilon^q$ is small enough.

Therefore, $\mathcal{G}(\sigma) \in \Sigma(\rho)$, and the operator $\mathcal{G}$ is well defined.

Finally we prove that $\mathcal{G}$ is a contraction. Let z_1 and z_2 be two functions that belong to $\Sigma(\rho)$:

$$\begin{aligned} |(\mathcal{G}(z_1) - \mathcal{G}(z_2))(t, s, Y)| &\le \Big|M(t + s)\Big[M^{-1}(s)\Big(c(z_1)(s, Y) - c(z_2)(s, Y)\Big) \\ &\quad + \int_0^t M^{-1}(\sigma + s)\Big(b(z_1)(\sigma, s, Y) - b(z_2)(\sigma, s, Y)\Big)\, d\sigma\Big]\Big|. \end{aligned} \tag{4.22}$$

We observe that, since the operator $\triangle_{2\pi\varepsilon}^{-1}$ is linear

$$M^{-1}(s)c(z_1) - M^{-1}(s)c(z_2) = \triangle_{2\pi\varepsilon}^{-1}(g(z_1) - g(z_2)).$$

By Lemma 4.4 we have

$$\|M^{-1}(s)c(z_1) - M^{-1}(s)c(z_2)\| \leq C_{D_0}\frac{e^{2\pi\varepsilon/\tau_2}}{2\pi\varepsilon}\|g(z_1) - g(z_2)\|. \tag{4.23}$$

Now we bound $\|b(z_1) - b(z_2)\|$. Until the end of the proof, z_0, z_1 and z_2 will stand for $z_0(t+s,Y)$, $z_1(t,s,Y)$ and $z_2(t,s,Y)$ respectively.

It is clear that we can write $b(z_1) - b(z_2)$ as

$$\begin{aligned} b(z_1) - b(z_2) &= \frac{1}{\mu\varepsilon^q}[X_0(z_0 + \mu\varepsilon^q z_1) - X_0(z_0 + \mu\varepsilon^q z_2) - \mu\varepsilon^q DX_0(z_0)(z_1 - z_2)] \\ &\quad + X_1(z_0 + \mu\varepsilon^q z_1, t/\varepsilon) - X_1(z_0 + \mu\varepsilon^q z_2, t/\varepsilon) \\ &= \frac{1}{\mu\varepsilon^q}(X_0(z_0 + \mu\varepsilon^q z_1) - X_0(z_0 + \mu\varepsilon^q z_2)) \\ &\quad - DX_0(z_0 + \mu\varepsilon^q z_1)(z_1 - z_2) \\ &\quad + DX_0(z_0 + \mu\varepsilon^q z_1)(z_1 - z_2) - DX_0(z_0)(z_1 - z_2) \\ &\quad + X_1(z_0 + \mu\varepsilon^q z_1, t/\varepsilon) - X_1(z_0 + \mu\varepsilon^q z_2, t/\varepsilon). \end{aligned}$$

Using the bounds of Lemma 4.1 we get

$$\begin{aligned} \|b(z_1) - b(z_2)\| &\leq |\mu|\varepsilon^q K(\|z_1 - z_2\|^2 + \|z_1\|\,\|z_1 - z_2\|) \\ &\quad + |\mu|\varepsilon^q K(\|z_1 - z_2\|^2 + \|z_1\|\,\|z_1 - z_2\|) \\ &= K|\mu|\varepsilon^q\|z_1 - z_2\|. \end{aligned} \tag{4.24}$$

Moreover, it is clear that

$$\|g(z_1) - g(z_2)\| \leq C_M 2\pi\varepsilon\|b(z_1) - b(z_2)\|.$$

Then, using (4.23) and (4.24) in (4.22), we obtain

$$\begin{aligned} \|\mathcal{G}(z_1) - \mathcal{G}(z_2)\| &\leq C_M C_{D_0}\frac{e^{2\pi\varepsilon/\tau_2}}{2\pi\varepsilon}\|g(z_1) - g(z_2)\| + 4\pi\varepsilon C_M^2\|b(z_1) - b(z_2)\| \\ &\leq K(D_0, M)\|b(z_1) - b(z_2)\| \\ &\leq K(D_0, M)|\mu|\varepsilon^q\|z_1 - z_2\| \\ &\leq \frac{1}{2}\|z_1 - z_2\| \end{aligned}$$

if $|\mu|\varepsilon^q$ is small enough.

Therefore, since $\mathcal{G}$ is a contraction, by the fixed point theorem, there exists a unique $z_1 \in \Sigma(\rho)$ such that $z_0(t+s,Y) + \mu\varepsilon^q z_1(t,s,Y)$ satisfies the conclusions of the proposition, except that z is C^0. Since z satisfies the equation

$$z(t,s,Y) = z(0,s,Y) + \int_0^t X_\mu(z(\sigma,s,Y),\sigma/\varepsilon)\,d\sigma,$$

it is C^1, and the proposition holds. □

4.5.4. Proof of Theorem 4.1. The proof of Theorem 4.1 has two parts. The first one consists on constructing flow box coordinates in $D_0 \times W$ using Proposition 4.1. The change of coordinates so obtained may be non-canonical. In the second step we modify these flow box coordinates in such way that they become canonical.

We begin by defining

$$w(u,\theta,Y) = z(\varepsilon\theta, u - \varepsilon\theta, Y).$$

Note that w is C^1 and analytic with respect to its first and third variables for $(u,Y) \in D_0(\kappa_1^\pm - \tau, \kappa_2, \kappa_3) \times W(r_1)$. Moreover, since the solutions of system (4.3) satisfy that $z(t+2\pi\varepsilon, s, Y) = z(t, s+2\pi\varepsilon, Y)$, we have that w is 2π-periodic respect to its second variable. This is a very important property because it allows us to extend the domain of w with respect to the θ variable, that is, we can consider w in the domain

$$D_0(\kappa_1^\pm - \tau, \kappa_2, 2\kappa_3) \times \mathbb{R} \times W(2r_1).$$

We have that

$$(w(t+s, t/\varepsilon, Y), t/\varepsilon) \tag{4.25}$$

is a new parameterization of the solutions of (4.3). Indeed,

$$\begin{aligned} \partial_t[w(t+s,t/\varepsilon,Y)] &= \partial_u w(t+s,t/\varepsilon,Y) + \frac{1}{\varepsilon}\partial_\theta w(t+s,t/\varepsilon,Y) \\ &= \partial_s z(t,s,Y) + [\partial_t z(t,s,Y)\varepsilon + \partial_s z(t,s,Y)(-\varepsilon)](1/\varepsilon) \\ &= \partial_t z(t,s,Y). \end{aligned}$$

We observe that, if $\mu = 0$, $w(u,\theta,Y) \equiv w_0(u,Y) = z_0(u,Y)$. We denote $z_1(\varepsilon\theta, u - \varepsilon\theta, Y)$ by $w_1(u,\theta,Y)$ and hence

$$w(u,\theta,Y) = w_0(u,Y) + \mu\varepsilon^q w_1(u,\theta,Y).$$

In the previous arguments we have not mentioned explicitly the dependence on the parameters, but it is clear that the continuity on ε and the analyticity on μ is maintained, and in particular w is C^0 in ε and analytic in μ.

LEMMA 4.5. *Let $\kappa_1^- > 2\kappa_0$. Under the hypotheses **H1,H2**, for any $\kappa_1^+ > \kappa_1^-$, $\kappa_2 > 0$, $\kappa_3 > 0$ such that $\kappa_1^- - 3\kappa_3 > \kappa_0$ there exist $r > 0$ small enough and two unique functions $\mathcal{T}$ and $\mathcal{Y}$ defined in $U(\kappa_1^\pm, \kappa_2, 0, r)$ such that*

$$w(\mathcal{T}(x,v,\theta), \theta, \mathcal{Y}(x,v,\theta)) = (x,v). \tag{4.26}$$

The functions $\mathcal{T}$ and $\mathcal{Y}$ are C^1, analytic in the (x,v) variables and 2π-periodic in θ. Moreover

$$\mathcal{T}(x,y,\theta) = \mathcal{T}_0(x,y) + O(\mu\varepsilon^q), \qquad \mathcal{Y}(x,y,\theta) = \mathcal{Y}_0(x,y) + O(\mu\varepsilon^q)$$

where $\mathcal{T}_0$ and $\mathcal{Y}_0$ are defined in (4.6) and (4.7).

PROOF. We define the function

$$G(S,Y,x,v,\theta,\mu,\varepsilon) = w(S,\theta,Y,\mu,\varepsilon) - (x,v).$$

on the set

$$D_0(\kappa_1^\pm - \tau, \kappa_2, 2\kappa_3) \times W(2r_1) \times U_0(\kappa_1^\pm, \kappa_2, 0, r) \times \mathbb{R} \times P$$

where

$$P = \{(\mu,\varepsilon) \in \mathbb{C} \times \mathbb{R} : |\mu| < \mu_0 \text{ and } 0 < \varepsilon < \varepsilon_0\}$$

with μ_0 and ε_0 small enough. Here we put explicitly the dependence on μ and ε of the solutions.

By the definitions of $\mathcal{T}_0$ and $\mathcal{Y}_0$ in (4.6) and (4.7) we have that, when $\mu = 0$,

$$z_0(\mathcal{T}_0(x,v), \mathcal{Y}_0(x,v)) = \psi_0(\mathcal{T}_0(x,v), \delta, \mathcal{Y}_0(x,v)) = (x,v) \tag{4.27}$$

where ψ_0 is introduced in Section 4.4. Then

$$\begin{aligned} G(\mathcal{T}_0(x,v),\mathcal{Y}_0(x,v),x,v,\theta,0,\varepsilon) &= w(\mathcal{T}_0(x,v),\theta,\mathcal{Y}_0(x,v),0,\varepsilon)-(x,v) \\ &= z_0(\mathcal{T}_0(x,v),\mathcal{Y}_0(x,v))-(x,v) \\ &= 0. \end{aligned}$$

We observe that, by Proposition 4.1, G is analytic on $(S,Y)\in D_0(\kappa_1^{\pm}-\tau,\kappa_2,2\kappa_3)\times W(2r_1)$.

Next we study the matrix $D_{S,Y}G$. Since the unperturbed system is Hamiltonian with Hamiltonian

$$\mathcal{F}_0(x,v)=\frac{v^2}{2}+vf(x),$$

the solution $z_0(u,Y)=(z_0^1(u,Y),z_0^2(u,Y))$ satisfies

$$\frac{Y^2}{2}+Yf(\delta)=\frac{(z_0^2(u,Y))^2}{2}+z_0^2(u,Y)f(z_0^1(u,Y)). \tag{4.28}$$

Differentiating with respect to Y in (4.28) we obtain

$$\begin{aligned} &Y+f(\delta) \\ &=(z_0^2(u,Y)+f(z_0^1(u,Y)))\partial_Y z_0^2(u,Y)+z_0^2(u,Y)f'(z_0^1(u,Y))\partial_Y z_0^1(u,Y). \end{aligned} \tag{4.29}$$

Evaluating (4.29) at $(u,Y)=(\mathcal{T}_0(x,v),\mathcal{Y}_0(x,v))$, we get

$$\mathcal{Y}_0+f(\delta)=(v+f(x))\partial_Y z_0^2(\mathcal{T}_0,\mathcal{Y}_0)+vf'(x)\partial_Y z_0^1(\mathcal{T}_0,\mathcal{Y}_0).$$

Now we prove that the derivative $\partial_{SY}G$ at $\mu=0$ is invertible. Using (4.27) and (4.29), the determinant of $D_{S,Y}G$ evaluated at $(S,Y)=(\mathcal{T}_0(x,v),\mathcal{Y}_0(x,v))$ is

$$\begin{aligned} \det(D_{SY}G) &= \partial_S z_0^1(\mathcal{T}_0,\mathcal{Y}_0)\partial_Y z_0^2(\mathcal{T}_0,\mathcal{Y}_0)-\partial_Y z_0^1(\mathcal{T}_0,\mathcal{Y}_0)\partial_S z_0^2(\mathcal{T}_0,\mathcal{Y}_0) \\ &= X_0^1(x,v)\partial_Y z_0^2(\mathcal{T}_0,\mathcal{Y}_0)-X_0^2(x,v)\partial_Y z_0^1(\mathcal{T}_0,\mathcal{Y}_0) \\ &= \mathcal{Y}_0+f(\delta) \end{aligned}$$

and by definition (4.7) of $\mathcal{Y}_0$, we obtain

$$\det(D_{SY}G)=-\sqrt{f^2(\delta)+v^2+2vf(x)}.$$

We recall that $|v|<r<r_0$, and that $C_0\delta^\beta\le|f(\delta)|\le C_1\delta^\beta$, hence, by Definition 4.1, taking $\delta_0'=\delta$

$$\begin{aligned} |\det(D_{SY}G)|^2 &\ge f(\delta)^2-|v|^2-2|vf(x)| \\ &\ge C_0^2\delta^{2\beta}-r_0^2-2r_0C_1\delta_0^\beta \\ &> 0. \end{aligned}$$

At this point it would be natural to apply the implicit function theorem to the equation

$$G(S,Y,x,v,\theta,\mu,\varepsilon)=0.$$

However to have a good control on the domains in which we will find the solution $\mathcal{T}$, $\mathcal{Y}$ in terms of $(x,v,\theta,\mu,\varepsilon)$ we follow the proof of the implicit function theorem using the special structure of the equation we deal with. We will work in a space of functions of the form

$$(S,Y)=h(x,v,\theta,\mu,\varepsilon).$$

In the rest of this proof we take the norm

$$\|(\xi,\eta)\| = \max\{|\xi|,|\eta|\}$$

for $(\xi,\eta)\in\mathbb{C}^2$. We define the space Γ of functions $h : U_0\times\mathbb{R}\times P\to\mathbb{C}^2$ which satisfy (we call $(x,v,\theta,\mu,\varepsilon)$ the variables of h)

(a) h is C^0.
(b) h is analytic in $(x,v,\mu)\in U_0(\kappa_1^{\pm},\kappa_2,0,r)\times\{\mu\in\mathbb{C} : |\mu|\le\mu_0\}$.
(c) h is C^1 and 2π-periodic with respect to θ.
(d) The norm

$$\begin{aligned}\|h\|_\Gamma &= \sup_{U_0\times\mathbb{R}\times P}\|h(x,v,\theta,\mu,\varepsilon)\| + \sup_{U_0\times\mathbb{R}\times P}\|\partial_\theta h(x,v,\theta,\mu,\varepsilon)\| \\ &\equiv \|h\|_\infty + \|\partial_\theta h\|_\infty\end{aligned}$$

is bounded.

We endow Γ with the norm $\|\cdot\|_\Gamma$ and it becomes a Banach space. We call $\Gamma(\rho)$ the closed ball of radius ρ of Γ, centered at $(\mathcal{T}_0(x,v),\mathcal{Y}_0(x,v))\in\Gamma$. We observe that, since $(\mathcal{T}_0(x,v),\mathcal{Y}_0(x,v))$ does not depend on θ, for any $h\in\Gamma(\rho)$,

$$\|h-(\mathcal{T}_0,\mathcal{Y}_0)\|_\Gamma = \|h-(\mathcal{T}_0,\mathcal{Y}_0)\|_\infty + \|\partial_\theta h\|_\infty.$$

We define the operator $\mathcal{G} : \Gamma(\rho)\to\Gamma(\rho)$ by

$$\mathcal{G}(h)(x,v,\theta,\mu,\varepsilon) = h - (D_{S,Y}G(\mathcal{T}_0,\mathcal{Y}_0,x,v,\theta,0,\varepsilon))^{-1}G(h,x,v,\theta,\mu,\varepsilon),$$

where in the right hand side, $h = h(x,v,\theta,\mu,\varepsilon)$, $\mathcal{T}_0 = \mathcal{T}_0(x,v)$ and $\mathcal{Y}_0 = \mathcal{Y}_0(x,v)$. $\mathcal{G}$ is well defined. Indeed, let $h\in\Gamma(\rho)$ with ρ small enough. By Section 4.5.1

$$(\mathcal{T}_0,\mathcal{Y}_0)\in D_0(\kappa_1^{\pm}-\tau,\kappa_2,\kappa_3)\times W(r_1),$$

thus, if ρ is small enough, $h\in D_0(\kappa_1^{\pm}-\tau,\kappa_2,2\kappa_3)\times W(2r_1)$ and then $\mathcal{G}(h)\in\Gamma$. Next we will check that $\mathcal{G}(h)\in\Gamma(\rho)$.

To shorten the notation we will not write the dependence on the variables (x,v,θ,ε), and we will denote $(\mathcal{T}_0,\mathcal{Y}_0)$ by h_0. By Taylor's theorem,

$$\begin{aligned}\mathcal{G}(h)(\mu) &= h - (D_{S,Y}G(h_0,0))^{-1}G(h,\mu) \\ &= h - (D_{S,Y}G(h_0,0))^{-1}\Big(G(h_0,0) + DG(h_0,0)(h-h_0,\mu)^T \\ &\quad + \int_0^1 (DG(\mathcal{Z}(\zeta)) - DG(h_0,0))(h-h_0,\mu)^T\,d\zeta\Big)\end{aligned}$$

where $DG\equiv(\partial_S G,\partial_Y G,\partial_\mu G)$ and $\mathcal{Z}(\zeta) = (h_0+\zeta(h-h_0),\zeta\mu)$. We observe that G is well defined in $\mathcal{Z}(\zeta)$ for all $\zeta\in[0,1]$. Then, using that

$$G(h_0,0) = G(\mathcal{T}_0,\mathcal{Y}_0,0) = 0,$$

we obtain

$$\begin{aligned}\mathcal{G}(h)(\mu) &= h_0 - (D_{S,Y}G(h_0,0))^{-1}\Big(\mu\partial_\mu G(h_0,0) \\ &\quad - \int_0^1 (DG(\mathcal{Z}(\zeta)) - DG(h_0,0))(h-h_0,\mu)^T\,d\zeta\Big).\end{aligned}$$

We observe that

$$\partial_\mu G(h_0,\mu) = \varepsilon^q w_1(\mathcal{T}_0,\theta,\mathcal{Y}_0,\mu,\varepsilon) + \mu\varepsilon^q\partial_\mu w_1(\mathcal{T}_0,\theta,\mathcal{Y}_0,\mu,\varepsilon). \tag{4.30}$$

Since G is analytic in h, G has its second derivative with respect to h bounded in $D_0(\kappa_1^\pm - \tau, \kappa_2, 2\kappa_3) \times W(2r_1)$, restricting κ_3 if necessary. Therefore

$$\|\mathcal{G}(h) - h_0\|_\infty \leq K(|\mu|\varepsilon^q + \rho^2 + \rho|\mu|) \leq \rho/2$$

if ρ and $|\mu|\varepsilon^q$ are small enough. Here we have used that $\|\partial_\mu G(h_0, 0)\|_\infty = O(\varepsilon^q)$ is bounded and that, by the mean value theorem,

$$\|\left[D_{S,Y}G(\mathcal{Z}(\zeta)) - D_{S,Y}G(h_0, 0)\right](h - h_0)\|_\infty \leq K\rho(\rho + |\mu|).$$

Using (4.30), we have that

$$\partial_{\theta\mu} G = O(\varepsilon^q).$$

Moreover, since h_0 does not depend on θ, and that

$$\partial_\theta w(S, \theta, Y) = O(\mu\varepsilon^q)$$

and consequently $\partial_\theta D_{SY} G = O(\mu\varepsilon^q)$, we have that

$$\begin{aligned}
\|\partial_\theta \mathcal{G}(h)(\mu)\| &\leq |\mu| K \int_0^1 |\partial_\theta (DG(\mathcal{Z}(\zeta)) - DG(h_0, 0))(h - h_0, \mu)^T| \, d\zeta \\
&\leq |\mu| \Big(\int_0^1 \zeta |\partial_\theta DG(\mathcal{Z}(\zeta))(\partial_\theta h, 0)^T (h - h_0, \mu)^T| \, d\zeta \\
&\quad + \int_0^1 |(DG(\mathcal{Z}(\zeta)) - DG(h_0, 0))(\partial_\theta h, 0)^T| \, d\zeta \Big) \\
&\leq |\mu| K (|\mu|\varepsilon^q \rho^2 + |\mu|\varepsilon^q + \rho^3) \\
&\leq \rho/2
\end{aligned}$$

if $|\mu|\varepsilon^q$ is small enough. In fact we can take $\rho = O(\mu\varepsilon^q)$.

The operator $\mathcal{G}$ is a contraction, thus the fixed point theorem can be applied, and we find functions $\mathcal{T}$ and $\mathcal{Y}$ such that for any $(x, v, \theta, \mu, \varepsilon) \in U(\kappa_1^\pm, \kappa_2, 0, r) \times P$,

$$w(\mathcal{T}, \theta, \mathcal{Y}, \mu, \varepsilon) = (x, v).$$

□

Now we prove that the flow can be straightened in $U(\kappa_1^\pm, \kappa_2, 0, r)$.

PROPOSITION 4.2. *Let $\kappa_1^- > 2\kappa_0$. If the hypotheses **H1**, **H2** hold, for any $\kappa_1^+ > \kappa_1^-$, $\kappa_2 > 0$, there exist $r > 0$ and a change of variables*

$$(x, v, \theta = \frac{t}{\varepsilon}) \in U(\kappa_1^\pm, \kappa_2, 0, r) \mapsto (T, F, \theta) = (\mathcal{T}(x, v, \theta), \mathcal{F}(x, v, \theta), \theta) \in \tilde{V}$$

analytic in the x, v variables, C^1 and 2π-periodic in θ, such that it transforms system (4.3) to

$$\begin{aligned}
\dot{T} &= 1 \\
\dot{F} &= 0 \\
\dot{\theta} &= 1/\varepsilon
\end{aligned}$$

and satisfies $\mathcal{T}(x, v, \theta) = \mathcal{T}_0(x, v) + O(\mu\varepsilon^q)$, $\mathcal{F}(x, v, \theta) = \mathcal{F}_0(x, v) + O(\mu\varepsilon^q)$ where $(x, v) \mapsto (\mathcal{T}_0(x, v), \mathcal{F}_0(x, v))$ is the corresponding change for the unperturbed system and is given in (4.6) and (4.4).

PROOF. We fix $(x,v)\in U_0(\kappa_1^\pm,\kappa_2,0,r)$ and we consider the solution $\psi(t)$ of system (4.3) such that $\psi(0)=(x,v,0)$. By Lemma 4.5, there exist $\mathcal{T}(x,v,0)$ and $\mathcal{Y}(x,v,0)$ such that

$$w(\mathcal{T}(x,v,0),0,\mathcal{Y}(x,v,0))=(x,v).$$

Moreover since the solutions of (4.3) can be parameterized as (4.25), taking $s=\mathcal{T}$ and $Y=\mathcal{Y}$ in (4.25) we obtain that

$$\tilde{\psi}(t)\equiv(w(\mathcal{T}(x,v,0)+t,t/\varepsilon,\mathcal{Y}(x,v,0)),t/\varepsilon), \tag{4.31}$$

is also a solution of (4.3) such that $\tilde{\psi}(0)=(x,v,0)=\psi(0)$. By uniqueness, $\tilde{\psi}=\psi$. On the other hand, if t is such that $\psi(t)\in U(\kappa_1^\pm,\kappa_2,0,r)$, by Lemma 4.5, applying (4.26) with $(x,v,\theta)=\psi(t)$, we obtain

$$\psi(t)=(w(\mathcal{T}(\psi(t)),t/\varepsilon,\mathcal{Y}(\psi(t))),t/\varepsilon). \tag{4.32}$$

Therefore (4.31) and (4.32) give us two expressions for the same solution $\psi(t)$. We observe that

$$\mathcal{T}_0(x,v)+t=\mathcal{T}_0(\psi_0(t))$$

therefore, by the uniqueness of the functions $\mathcal{T}$ and $\mathcal{Y}$ given in Lemma 4.5, we have

$$\begin{aligned}\mathcal{T}(\psi(t)) &= \mathcal{T}(x,v,0)+t\\ \mathcal{Y}(\psi(t)) &= \mathcal{Y}(x,v,0)\end{aligned} \tag{4.33}$$

and then

$$\begin{aligned}\frac{d}{dt}\mathcal{T}(\psi(t)) &= 1\\ \frac{d}{dt}\mathcal{Y}(\psi(t)) &= 0.\end{aligned}$$

We define a new function

$$\mathcal{F}(x,v,\theta)=\mathcal{F}_0(\delta,\mathcal{Y}(x,v,\theta))$$

where $\mathcal{F}_0$ is the Hamiltonian of the unperturbed system given in (4.4). We recall that

$$\|\mathcal{Y}-\mathcal{Y}_0\|_\infty=O(\mu\varepsilon^q),$$

then, since $\mathcal{F}_0$ is constant along the trajectories of the unperturbed system,

$$\begin{aligned}\mathcal{F}(x,v,\theta) &= \mathcal{F}_0(\delta,\mathcal{Y}_0(x,v))+O(\mu\varepsilon^q)\\ &= \mathcal{F}_0(x,v)+O(\mu\varepsilon^q).\end{aligned}$$

Therefore, from (4.33), it is easily seen that

$$(T,F,\theta)=(\mathcal{T}(x,v,\theta),\mathcal{F}(x,v,\theta),\theta)$$

transforms (4.3) in $U(\kappa_1^\pm,\kappa_2,0,r)$ to

$$\begin{aligned}\dot{T} &= 1\\ \dot{F} &= 0\\ \dot{\theta} &= 1/\varepsilon\end{aligned}$$

and the statement holds. □

Now we turn to modify the change of variables to get a canonical one. Before starting the result we need some preliminary calculations.

We denote by $\psi(t, x, v)$ the solution of system (4.3) such that $(x, v, 0) = \psi(0)$. In the proof of Proposition 4.2, concretely in (4.33), we have seen that

$$\begin{aligned} \mathcal{T}(\psi(t)) &= \mathcal{T}(x, v, 0) + t \\ \mathcal{F}(\psi(t)) &= \mathcal{F}(x, v, 0). \end{aligned} \tag{4.34}$$

We introduce the matrix

$$\Phi(t) = \begin{pmatrix} \partial_x \mathcal{T}(\psi(t)) & \partial_v \mathcal{T}(\psi(t)) \\ \partial_x \mathcal{F}(\psi(t)) & \partial_v \mathcal{F}(\psi(t)) \end{pmatrix}.$$

Differentiating with respect to (x, v) in both sides of (4.34), we obtain

$$\Phi(t) \begin{pmatrix} \partial_x \psi_1(t) & \partial_v \psi_1(t) \\ \partial_x \psi_2(t) & \partial_v \psi_2(t) \end{pmatrix} = \Phi(0).$$

Since system (4.3) is Hamiltonian,

$$\det \begin{pmatrix} \partial_x \psi_1(t) & \partial_v \psi_1(t) \\ \partial_x \psi_2(t) & \partial_v \psi_2(t) \end{pmatrix} = 1$$

and therefore

$$\det \Phi(t) = \det \Phi(0) \tag{4.35}$$

for all t for which the solution is defined. Moreover, we know that for $\mu = 0$, $\det \Phi(t) = \det \Phi(0) = 1$, thus

$$\det \Phi(t) = 1 + \mu \varepsilon^q \tilde{g}(\psi(t))$$

where $\tilde{g} = \tilde{g}(x, v, \theta)$ is some C^1 function, analytic in (x, v) and 2π-periodic in θ. Moreover from (4.35) it is clear that

$$\frac{d}{dt} \tilde{g}(\psi(t)) = 0. \tag{4.36}$$

We define the function $g : U_g \to \mathbb{C}$, by

$$g(T, F, \theta) = \tilde{g}(w(T, \theta, f(\delta) - \sqrt{f^2(\delta) + 2F}), \theta).$$

The function g is C^1, analytic in (T, F) and 2π-periodic in θ. If we differentiate with respect to the time, t, in g evaluated on the solutions of $\dot{T} = 1$, $\dot{F} = 0$, $\dot{\theta} = 1/\varepsilon$, by (4.36), we get the following equality:

$$0 = \partial_T g + \frac{1}{\varepsilon} \partial_\theta g. \tag{4.37}$$

To deal with equation (4.37), we define the change $(\xi, \eta) = (T + \varepsilon\theta, T - \varepsilon\theta)$ and the function

$$h(\xi, \eta, F) = g((\xi + \eta)/2, F, (\xi - \eta)/2\varepsilon)$$

defined in

$$\{(\xi, \eta, F) \in \mathbb{C}^3 : ((\xi + \eta)/2, F, (\xi - \eta)/2\varepsilon) \in U_g\}.$$

Then, by (4.37)

$$\partial_\xi h = 0$$

therefore,

$$g(T,F,\theta) = h(T+\varepsilon\theta, T-\varepsilon\theta, F) = h(0,T-\varepsilon\theta,F)$$

is a function that only depends on first integrals of system (4.3). We define the function $\rho(F,S)$ by the condition

$$1+\partial_F\rho(F,S) = \frac{1}{1+\mu\varepsilon^q h(0,S,F)}.$$

We remark that, since h is analytic in F and S, the function ρ is $O(\mu\varepsilon^q)$.

PROPOSITION 4.3. *The change of variables defined by*

$$(x,v,\theta=t/\varepsilon)\in U(\kappa_1^{\pm},\kappa_2,0,r)\to(T,I,\theta)=(\mathcal{T}(x,v,\theta),\mathcal{I}(x,v,\theta),\theta)\in V$$

with $\mathcal{I}(x,v,\theta)=\mathcal{F}(x,v,\theta)+\rho(\mathcal{F}(x,v,\theta),\mathcal{T}(x,v,\theta)-\theta\varepsilon)$ *is canonical and is such that transforms system (4.3) to*

$$\begin{aligned}\dot T &= 1\\ \dot I &= 0\\ \dot\theta &= 1/\varepsilon.\end{aligned}$$

Moreover, $\mathcal{T}(x,v,\theta)=\mathcal{T}_0(x,v)+O(\mu\varepsilon^q)$ *and* $\mathcal{I}(x,v,\theta)=\mathcal{F}_0(x,v)+O(\mu\varepsilon^q)$ *where* $\mathcal{T}_0$ *and* $\mathcal{F}_0$ *is the corresponding change in the unperturbed case.*

PROOF. Let $\mathcal{I}=\mathcal{I}(x,v,\theta)$ be as in the statement. Along the solutions of (4.3),

$$\dot I = \dot F + D_F\rho(F,S)\dot F + D_S\rho(F,S)(\dot T-\varepsilon\dot\theta)=0.$$

Thus, this change transforms system (4.3) to $\dot T=1$, $\dot I=1$, $\dot\theta=1/\varepsilon$.

To see that the change is canonical we only have to calculate the determinant of

$$C(x,v,\theta)=\begin{pmatrix}\partial_x\mathcal{T} & \partial_v\mathcal{T}\\ \partial_x\mathcal{I} & \partial_v\mathcal{I}\end{pmatrix}.$$

We have

$$\begin{aligned}\det C(x,v,\theta) &= \partial_x\mathcal{T}\partial_v\mathcal{I}-\partial_v\mathcal{T}\partial_x\mathcal{I}\\ &= [\partial_v\mathcal{F}+\partial_F\rho(\mathcal{F},\mathcal{T}-\theta\varepsilon)\partial_v\mathcal{F}+\partial_S\rho(\mathcal{F},\mathcal{T}-\theta\varepsilon)\partial_v\mathcal{T}]\,\partial_x\mathcal{T}\\ &\quad -[\partial_x\mathcal{F}+\partial_F\rho(\mathcal{F},\mathcal{T}-\theta\varepsilon)\partial_x\mathcal{F}+\partial_S\rho(\mathcal{F},\mathcal{T}-\theta\varepsilon)\partial_x\mathcal{T}]\,\partial_v\mathcal{T}\\ &= (\partial_v\mathcal{F}\partial_x\mathcal{T}-\partial_x\mathcal{F}\partial_v\mathcal{T})(1+\partial_F\rho(\mathcal{F},\mathcal{T}-\theta\varepsilon))\\ &= (1+\mu\varepsilon^q h(0,\mathcal{T}-\theta\varepsilon,\mathcal{F}))(1+\partial_F\rho(\mathcal{F},\mathcal{T}-\theta\varepsilon))\\ &= 1.\end{aligned}$$

□

Now we turn to the system in the original variables (x,y,θ). We define

$$\begin{aligned}\mathcal{T}^1(x,y,\theta) &= \mathcal{T}(C(x,y,\theta))\\ \mathcal{I}^1(x,y,\theta) &= \mathcal{I}(C(x,y,\theta))\end{aligned}$$

where C defined in (4.2). It is clear that the change

$$(x,y,\theta)\in V(\kappa_1^{\pm},\kappa_2,0,r)\mapsto(T,I,\theta)=(\mathcal{T}^1(x,y,\theta),\mathcal{I}^1(x,y,\theta),\theta)\in\mathcal{V}$$

is canonical, since it is the composition of two canonical changes. Moreover

$$\begin{aligned} \dot{T} &= 1 \\ \dot{I} &= 0 \end{aligned}$$

and

$$\begin{aligned} \mathcal{T}^1(x,y,\theta) &= \mathcal{T}_0(x,y-f(x)) + O(\mu\varepsilon^q) \\ \mathcal{I}^1(x,v,\theta) &= \mathcal{I}_0(x,y-f(x)) + O(\mu\varepsilon^q) \\ &= \mathcal{F}_0(x,y-f(x)) + O(\mu\varepsilon^q) \end{aligned}$$

where $\mathcal{T}_0(x,y-f(x))$ and $\mathcal{F}_0(x,y-f(x)) = h_0(x,y)$ is the change when $\mu = 0$ in the original variables. This ends the proof of Theorem 4.1.

4.6. Proof of Theorem 4.2

We consider the averaged system obtained in Proposition 3.1 (Chapter 3) whose equation is:

$$(4.38)\qquad \begin{aligned} \dot{x} &= y + \mu\varepsilon^{p+3}\partial_y F_{2n-2}(x,y,\theta) + \mu^2\varepsilon^{p+2}\partial_y R_{2k-2}(x,y) \\ \dot{y} &= -V'(x) - \mu\varepsilon^{p+3}\partial_x F_{2n-2}(x,y,\theta) - \mu^2\varepsilon^{p+2}\partial_x R_{2k-2}(x,y) \\ \dot{\theta} &= 1/\varepsilon. \end{aligned}$$

We denote by

$$\tilde{\gamma}^{\mathrm{s}}_{\mu,\varepsilon}(t,s) = (\tilde{\alpha}^{\mathrm{s}}_{\mu,\varepsilon}(t,s), \tilde{\beta}^{\mathrm{s}}_{\mu,\varepsilon}(t,s))$$

the stable curve of this system and by

$$\gamma_0(t+s) = (\alpha_0(t+s), \beta_0(t+s))$$

the homoclinic orbit of the unperturbed system.

We fix $T > 0$ big enough, then for $\kappa_1^- = T-1$, $\kappa_1^+ = 2T+1$ and $\kappa_2 = 3a/2$ we have that

$$\gamma_0(t+s) \in V_0(\kappa_1^{\pm} \mp 1, 2\kappa_2/3, 0, 0)$$

for $T \le t + \operatorname{Re} s \le 2T$ and $|\operatorname{Im} s| \le a$.

We observe that the averaged system (4.38) satisfies the hypotheses **H1**, **H2**, of Theorem 4.1 with $q = p+2$, therefore there exists $r < r_0$ small enough and a canonical change of variables, defined in the set $V(\kappa_1^{\pm}, \kappa_2, 0, r)$, which we denote by $(T,I) = (\mathcal{T}^1(\bar{x},\bar{y},\theta), \mathcal{I}^1(\bar{x},\bar{y},\theta))$, such that transforms system (4.38) to

$$\begin{aligned} \dot{T} &= 1 \\ \dot{I} &= 0. \end{aligned}$$

Moreover, $\mathcal{T}^1(\bar{x},\bar{y},\theta) = \mathcal{T}_0(\bar{x},\bar{y}) + O(\mu\varepsilon^{p+2})$ and $\mathcal{I}^1(\bar{x},\bar{y},\theta) = \mathcal{I}_0(\bar{x},\bar{y}) + O(\mu\varepsilon^{p+2})$.

We must see that the parametric representation of the stable manifold of system (4.38) enters the domain of analyticity of $(\mathcal{T}^1(x,y,\theta), \mathcal{I}^1(x,y,\theta))$.

Using the above definition, it is clear that, by continuity, for (t,s) such that $T \le t + \operatorname{Re} s \le 2T$ and $|\operatorname{Im} s| \le a$ we have that

$$\tilde{\gamma}^{\mathrm{s}}_{\mu,\varepsilon}(t,s) \in V_0(\kappa_1^{\pm}, \kappa_2, 0, r)$$

if $|\mu|$ is small enough.

We call $\mathcal{C}$ the change from the initial to the averaged system, defined in Proposition 3.1 (Chapter 3). We write $(x, y, \theta) = \mathcal{C}(\bar{x}, \bar{y}, \theta)$. Where $(\bar{x}, \bar{y})$ denote the variables of the averaged systems. Moreover, we know that

$$(x, y) = (\bar{x}, \bar{y}) + O(\mu\varepsilon^{p+1}).$$

We define new flow box coordinates

$$(\mathcal{T}^2(x, y, \theta), \mathcal{I}^2(x, y, \theta)) = (\mathcal{T}^1(\mathcal{C}^{-1}(x, y, \theta)), \mathcal{I}^1(\mathcal{C}^{-1}(x, y, \theta))).$$

Since the change $\mathcal{C}$ is $O(\mu\varepsilon^{p+1})$ close to the identity and canonical, the new change is also canonical and satisfies

$$(\mathcal{T}^2(x, y, \theta), \mathcal{I}^2(x, y, \theta)) = (\mathcal{T}_0^2(x, y), \mathcal{I}_0^2(x, y)) + O(\mu\varepsilon^{p+1}).$$

The change $\mathcal{T}_0^2(x, y), \mathcal{I}_0^2(x, y) = h_0(x, y)$ is the corresponding change for $\mu = 0$. The domain of the new change is $\mathcal{C}(V(\kappa_1^{\pm}, \kappa_2, 0, r))$. Moreover it is clear that the change $(T, I) = (\mathcal{T}^2(x, y, \theta), \mathcal{I}^2(x, y, \theta))$ transforms system (1.1) of Chapter 1 to

$$\begin{aligned} \dot{T} &= 1 \\ \dot{I} &= 0. \end{aligned}$$

Let $\gamma_{\mu,\varepsilon}^{\mathrm{s}}(t, s)$ be the parameterization of the stable manifold of system (1.1) of Chapter 1 given in Theorem 3.1. Since $\gamma_{\mu,\varepsilon}^{\mathrm{s}} = \mathcal{C}(\tilde{\gamma}_{\mu,\varepsilon}^{\mathrm{s}})$ we have that $\gamma_{\mu,\varepsilon}^{\mathrm{s}}$ belongs to the domain of the new flow box coordinates.

We define a new change of variables. Let s_0 be such that $|\operatorname{Im} s_0| \leq a$ and $t_0 \in \mathbb{R}$. We define

$$(x^*, y^*) = \gamma_{\mu,\varepsilon}^{\mathrm{s}}(T - \operatorname{Re} s_0, s_0)$$

and the parameter

$$\tau = s_0 - \mathcal{T}^2(x^*, y^*, (T - \operatorname{Re} s_0)/\varepsilon) - t_0.$$

We observe that, if $\mu = 0$, the constant τ is

$$\tau = \int_{x_0}^{\mathcal{C}^1(\delta, f(\delta), 0)} \frac{ds}{\sqrt{2h_0(x^*, y^*) - 2V(s)}} - t_0$$

where $\mathcal{C}^1$ denotes the first component of the change and x_0 is the initial condition of γ_0 as is defined in hypothesis **HP1**. τ does not depend on (x^*, y^*) while (x^*, y^*) belongs to the stable manifold of the unperturbed system.

We define the functions

$$\begin{aligned} \mathcal{S}(x, y, \theta) &= \mathcal{T}^2(x, y, \theta) + \tau \\ \mathcal{E}(x, y, \theta) &= \mathcal{I}^2(x, y, \theta) - \mathcal{I}^2(x^*, y^*, \theta). \end{aligned}$$

Next we will see that, since $\dot{\mathcal{T}}^2 = 1$ and $\dot{\mathcal{I}}^2 = 0$, we have that for $s = s_0$,

$$\mathcal{S}(\gamma_{\mu,\varepsilon}^{\mathrm{s}}(t, s_0), t/\varepsilon) = t - t_0 + s_0.$$

Indeed, using the definition of τ

$$\begin{aligned} \mathcal{S}(\gamma_{\mu,\varepsilon}^{\mathrm{s}}(t, s_0), t/\varepsilon) &= \mathcal{T}^2(\gamma_{\mu,\varepsilon}^{\mathrm{s}}(t, s_0), t/\varepsilon) + \tau \\ &= t + \mathcal{T}^2(\gamma_{\mu,\varepsilon}^{\mathrm{s}}(T - \operatorname{Re} s_0, s_0), (T - \operatorname{Re} s_0)/\varepsilon) + \tau \\ &= t + \mathcal{T}^2(x^*, y^*, (T - \operatorname{Re} s_0)/\varepsilon) + \tau \\ &= t - t_0 + s_0. \end{aligned}$$

Let $s \in \mathbb{C}$, $s \neq s_0$. Using that, for the unperturbed Hamiltonian system we have that $\mathcal{S}_0(\gamma_0(t+s)) = t - t_0 + s$ and the estimates $\mathcal{S}(x, y, \theta) = \mathcal{S}_0(x, v) + O(\mu\varepsilon^{p+1})$, $\gamma^{\mathrm{s}}_{\mu,\varepsilon}(t, s) = \gamma_0(t+s) + O(\mu\varepsilon^{p+1})$, we obtain

$$\mathcal{S}(\gamma^{\mathrm{s}}_{\mu,\varepsilon}(t, s), t/\varepsilon) = t - t_0 + s + \mu\varepsilon^{p+1}\mathcal{X}(s).$$

We note that we have $\mathcal{X}(s_0) = 0$, and that we have a lot of freedom to choose s_0 and t_0. Each choice gives a slightly different definition of $\mathcal{S}$.

We also note that $\mathcal{X}(s)$ is $2\pi\varepsilon$-periodic in s. Indeed, we have that

$$\begin{aligned} \mathcal{S}(\gamma^{\mathrm{s}}_{\mu,\varepsilon}(t, s + 2\pi\varepsilon), t/\varepsilon) &= t - t_0 + s + 2\pi\varepsilon + \mu\varepsilon^{p+1}\mathcal{X}(s + 2\pi\varepsilon) \\ \mathcal{S}(\gamma^{\mathrm{s}}_{\mu,\varepsilon}(t + 2\pi\varepsilon, s), t/\varepsilon) &= t - t_0 + 2\pi\varepsilon + s + \mu\varepsilon^{p+1}\mathcal{X}(s) \end{aligned}$$

and, since $\gamma^{\mathrm{s}}_{\mu,\varepsilon}(t, s + 2\pi\varepsilon) = \gamma^{\mathrm{s}}_{\mu,\varepsilon}(t + 2\pi\varepsilon, s)$, from the previous equations we obtain

$$\mathcal{X}(s + 2\pi\varepsilon) = \mathcal{X}(s).$$

Finally,

$$\begin{aligned} \mathcal{E}(\gamma^{\mathrm{s}}_{\mu,\varepsilon}(t, s), t/\varepsilon) &= \mathcal{I}^2(\gamma^{\mathrm{s}}_{\mu,\varepsilon}(t, s), t/\varepsilon) - \mathcal{I}^2(x^*, y^*, 0) \\ &= 0. \end{aligned}$$

This ends the proof of Theorem 4.2.

5. The Extension Theorem

This short Chapter is devoted to recall the statement of the extension theorem which is given in [**DS2**]. This theorem is stated for systems of the form

$$\begin{aligned} \dot{x} &= y + \mu\varepsilon^p \partial_y h_1(x, y, t/\varepsilon) \\ \dot{y} &= -V'(x) - \mu\varepsilon^p \partial_x h_1(x, y, t/\varepsilon) \end{aligned}$$

such that the unperturbed system has a homoclinic orbit, $\gamma_0(u) = (\alpha_0(u), \beta_0(u))$ and β_0 is an analytic function in $|\operatorname{Im} u| < a$ and has singularities at $u = \pm ia$ which are poles.

Following the proof in [**DS2**] one can see that one can replace the condition of $u = \pm ia$ being poles by $u = \pm ia$ being branching points in the sense we have introduced in **HP1** in Chapter 1. For this reason here we do not reproduce the proof of the extension theorem.

The goal of this theorem is to extend the domain of the parameterization $\gamma^{\mathrm{u}}_{\mu,\varepsilon}(t, s)$ of the unstable manifold (in our case produced in Chapter 3) until it enters the domain of the flow box coordinates. To do this, the parameterizations $\gamma^{\mathrm{u}}_{\mu,\varepsilon}(t, s)$ and $\gamma_0(t + s)$ are compared in the complex domain $D^{\mathrm{ext}}_{\varepsilon}$:

$$D^{\mathrm{ext}}_{\varepsilon} \equiv \{(t, s) \in \mathbb{R} \times \mathbb{C} : |t + \operatorname{Re} s| \leq 2T \ , \ |\operatorname{Im} s| \leq a - \varepsilon\}.$$

The extension theorem gives a useful bound for the distance between the unstable manifold $\gamma^{\mathrm{u}}_{\mu,\varepsilon}$ and the homoclinic orbit γ_0 of the unperturbed system for $(t, s) \in D^{\mathrm{ext}}_{\varepsilon}$. That is,

$$\gamma^{\mathrm{u}}_{\mu,\varepsilon}(t, s) - \gamma_0(t + s) = O(\mu\varepsilon^{\nu}),$$

where ν is a parameter which depends on the system.

Therefore, if $\nu \geq 0$ and μ and ε are small enough $\gamma^{\mathrm{u}}_{\mu,\varepsilon}(t, s)$, for some values of (t, s), belongs to the domain of the flow box coordinates.

The extension theorem is

THEOREM 5.1. *Let $z(t, s) = (x(t, s), y(t, s))$ be a family of solutions of*

$$\begin{aligned} \dot{x} &= y + \mu\varepsilon^p \partial_y h_1(x, y, t/\varepsilon, \mu, \varepsilon) \\ \dot{y} &= -V'(x) - \mu\varepsilon^p \partial_x h_1(x, y, t/\varepsilon, \mu, \varepsilon) \end{aligned}$$

defined for $t_0 + \operatorname{Re} s = -2T$, for some $T > 0$, such that

$$z(t_0, s) - \gamma_0(t_0 + s) - \mu\varepsilon^{p+1} G(\gamma_0(t_0 + s), t_0/\varepsilon, \mu, \varepsilon) = O(\mu\varepsilon^{p+2}),$$

where G is the function such that

$$\partial_\theta G(x, y, \theta, \mu, \varepsilon) = (\partial_y h_1(x, y, \theta, \mu, \varepsilon), -\partial_x h_1(x, y, \theta, \mu, \varepsilon))$$

and has zero mean with respect to θ, and $(t_0, s) \in D^{\mathrm{ext}}_{\varepsilon}$ verifies $t_0 + \operatorname{Re} s = -2T$.

Let ℓ be defined by (1.2). We assume that

$$\nu \equiv p - \ell \geq 0.$$

Then, there exist ε_0, μ_0 and K such that the solution $z(t, s)$ can be extended to values of $t \in [t_0, 2T - \operatorname{Re} s]$, with the bound

$$|z(t, s) - \gamma_0(t + s)| \leq K\mu\varepsilon^{p-\ell} \tag{5.1}$$

for $(t,s) \in D_\varepsilon^{\mathrm{ext}}$, $0 < \varepsilon \le \varepsilon_0$ *and* $|\mu| \le \mu_0$.

REMARK 5.1. *Theorem 5.1 is also valid if* V *is a trigonometric polynomial and we assume that* h_1 *is also a trigonometric polynomial in* x *and a polynomial in* y. *We observe that in this case* $\alpha_0(u) \sim iC\log(u \mp ia)$ *near of singularity* $u = \pm ia$ *with* C *a constant which depends on the degree of* V. *(See* [**DS2**] *for more details about this case).*

REMARK 5.2. *We note that, if* $s \in \mathbb{R}$, *the estimate (5.1) can be improved. Concretely, for* $(t,s) \in \mathbb{R}^2$ *such that* $-2T \le t+s \le 2T$ *we have that*

$$z(t_0,s) - \gamma_0(t_0+s) = O(\mu\varepsilon^{p+1}).$$

6. Splitting of separatrices

6.1. Introduction

This chapter is devoted to prove Theorem 1.1 and Corollary 1.1. To prove these results we will use the results established in the previous chapters. In particular the parameterization of the stable manifold, in Chapter 3, and the flow box coordinates developed in Chapter 4 will play an important role. Let

$$\begin{aligned} \dot{x} &= y + \mu\varepsilon^p \partial_y h_1(x, y, t/\varepsilon) \\ \dot{y} &= -V'(x) - \mu\varepsilon^p \partial_x h_1(x, y, t/\varepsilon). \end{aligned} \tag{6.1}$$

We recall that, in Chapter 4, we have constructed flow box coordinates $(S, E) = (\mathcal{S}(x, y, t/\varepsilon), \mathcal{E}(x, y, t/\varepsilon))$ defined in a complex neighborhood of a piece of the stable manifold of system (6.1). In these coordinates the original system becomes the simple equation:

$$\dot{S} = 1, \qquad \dot{E} = 0.$$

These variables evaluated on the parameterization of the stable manifold of system (6.1), $\gamma^{\mathrm{s}}_{\mu,\varepsilon}(t, s)$ (with s depending on the initial condition) take the values

$$\mathcal{S}(\gamma^{\mathrm{s}}_{\mu,\varepsilon}(t, s), t/\varepsilon) = t - t_0 + s + \mu\varepsilon^{p+1}\mathcal{X}(s), \qquad \mathcal{E}(\gamma^{\mathrm{s}}_{\mu,\varepsilon}(t, s), t/\varepsilon) = 0.$$

We can prove the existence of primary homoclinic points by using that the Poincaré map is area preserving and that system (6.1) is a perturbation of one which has a homoclinic connection. Moreover, by the extension theorem, we get that the parameterization of the unstable manifold $\gamma^{\mathrm{u}}_{\mu,\varepsilon}(t, s)$ enters, for some values of (t, s), into the domain of the flow box coordinates.

In flow box coordinates the stable manifold corresponds to $E = 0$. We will see that, in these variables, the unstable manifold can be written as $E = \phi(S)$, with ϕ a suitable function, therefore we have the following situation:

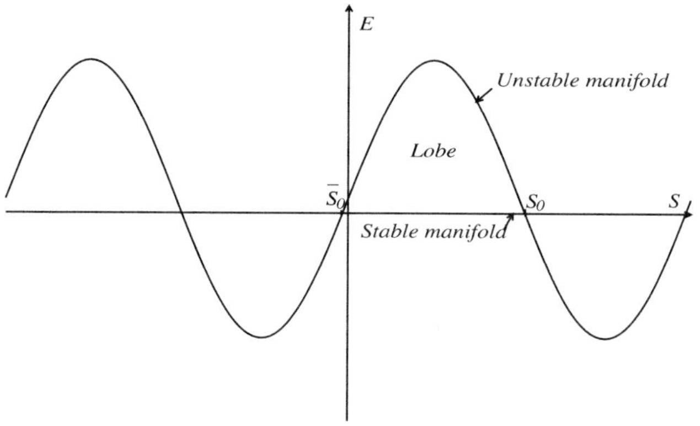

Consequently the area of the lobe generated by two consecutive intersections between the stable and the unstable manifold, expressed in flow box coordinates, is

$$A = \left| \int_{\bar{S}_0}^{S_0} \phi(S)\, dS \right|. \tag{6.2}$$

The function ϕ is called the splitting function. Since the change from the original variables to the flow box variables is canonical, the area given by (6.2) is the same as the area of the corresponding lobe in the original variables (x, y).

The scheme of the proof is the same as the one given in [**DS2**]. For the convenience of the reader we present the main points of it. In Section 6.2 we construct the splitting function and we establish its properties. In Section 6.3 we prove the main results. The flow box coordinates we use and the definition of the splitting function permit to prove that the Melnikov function is a good approximation of the splitting function.

Finally we will prove Corollary 1.1. We recall that we consider the case that the singularities of the homoclinic orbit may be branching points in the sense we have indicated in Chapter 1. This case is not considered in [**DS2**]. This fact forces several technicalities in the computation of the Melnikov function.

The asymptotic computations of the Melnikov function are deferred to Section 6.4.

6.2. The splitting function

To define the splitting function and to establish the properties we shall need we begin by recalling some notation and some previous results.

We recall the definitions of the sets

$$\begin{aligned} D_\varepsilon^{\mathrm{ext}} &= \{(t,s)\in\mathbb{R}\times\mathbb{C} : |t+\operatorname{Re}s|\le 2T,\ \ |\operatorname{Im}s|\le a-\varepsilon\},\\ D^{\mathrm{s}} &= \{(t,s)\in\mathbb{R}\times\mathbb{C} : t+\operatorname{Re}s\ge T,\ \ |\operatorname{Im}s|\le a\},\\ D^{\mathrm{u}} &= \{(t,s)\in\mathbb{R}\times\mathbb{C} : t+\operatorname{Re}s\le -T,\ \ |\operatorname{Im}s|\le a\}. \end{aligned}$$

The homoclinic orbit of the unperturbed system is

$$\gamma_0(u) = (\alpha_0(u), \beta_0(u))$$

which is defined in (at least in) $\{u\in\mathbb{C} : |\operatorname{Im}u| < a\}$.

We denote $\gamma_{\mu,\varepsilon}^{\mathrm{s,u}}(t,s)$ the parameterizations of the stable and unstable manifolds of the perturbed system and we recall that, by Theorem 3.1, the invariant curves are solutions with respect to t and satisfy

$$\gamma_{\mu,\varepsilon}^{\mathrm{s,u}}(t+2\pi\varepsilon, s) = \gamma_{\mu,\varepsilon}^{\mathrm{s,u}}(t, s+2\pi\varepsilon). \tag{6.3}$$

Moreover,

$$\begin{aligned} \gamma_{\mu,\varepsilon}^{\mathrm{s}}(t,s) &= \gamma_0(t+s) + O(\mu\varepsilon^{p+1}) \quad \text{for } (t,s)\in D^{\mathrm{s}}\\ \gamma_{\mu,\varepsilon}^{\mathrm{u}}(t,s) &= \gamma_0(t+s) + O(\mu\varepsilon^{p+1}) \quad \text{for } (t,s)\in D^{\mathrm{u}}. \end{aligned}$$

We call $U\equiv\mathcal{C}(V(\kappa_1^{\pm},\kappa_2,0,r))$ the domain of the flow box coordinates (S,E), constructed in Chapter 4 (see in particular Theorem 4.2), which is a neighborhood of

$$\{\gamma_0(t+s) : T\le t+\operatorname{Re}s\le 2T,\ \ |\operatorname{Im}s|\le a\}$$

and is independent of μ,ε. In fact, we have chosen $\kappa_1^{\pm}$, and κ_2 such that, $\gamma_0(t+s)\in V_0(\kappa_1^{\pm}\mp 1, 2\kappa_2/3, 0, 0)$.

By the extension theorem of Chapter 5, the domain of the parameterization of the unstable manifold $\gamma_{\mu,\varepsilon}^{\mathrm{u}}(t,s)$ can be extended to values of $(t,s)\in D_\varepsilon^{\mathrm{ext}}$, and in this extended set verifies

$$\gamma_{\mu,\varepsilon}^{\mathrm{u}}(t,s) = \gamma_0(t+s) + O(\mu\varepsilon^{\nu})$$

where $\nu = p - \ell$.

Since $\nu \geq 0$ and $\gamma_0(t+s) \in V_0(\kappa_1^\pm \mp 1, 2\kappa_2/3, 0, 0)$, there exist ε_0 and μ_0 such that, for all $0 < \varepsilon < \varepsilon_0$, $|\mu| < \mu_0$ and (t,s) such that $T \leq t + \operatorname{Re} s \leq 2T$ and $|\operatorname{Im} s| \leq a - \varepsilon$, the manifolds $\gamma^{\mathrm{u}}_{\mu,\varepsilon}(t,s)$ and $\gamma^{\mathrm{s}}_{\mu,\varepsilon}(t,s)$ are so close to $\gamma_0(t+s)$ that both belong to U for those values of (t,s).

The expressions $\mathcal{S}(\gamma^{\mathrm{u}}_{\mu,\varepsilon}(t,s), t/\varepsilon) - (t - t_0)$ and $\mathcal{E}(\gamma^{\mathrm{u}}_{\mu,\varepsilon}(t,s), t/\varepsilon)$ are well defined for $s \in \mathbb{C}$ such that $T \leq t + \operatorname{Re} s \leq 2T$ and $|\operatorname{Im} s| \leq a - \varepsilon$. Then, choosing t arbitrarily we can define

$$\mathcal{S}^{\mathrm{u}}(s) = \mathcal{S}(\gamma^{\mathrm{u}}_{\mu,\varepsilon}(t,s), t/\varepsilon) - (t - t_0), \qquad \mathcal{E}^{\mathrm{u}}(s) = \mathcal{E}(\gamma^{\mathrm{u}}_{\mu,\varepsilon}(t,s), t/\varepsilon). \tag{6.4}$$

Moreover, according to Theorem 4.2, they do not depend on time. We choose it in such a way that $T \leq t + \operatorname{Re} s \leq 2T$.

LEMMA 6.1. *The functions $\mathcal{S}^{\mathrm{u}}$ and $\mathcal{E}^{\mathrm{u}}$ satisfy the following properties*

a) *The functions $\mathcal{S}^{\mathrm{u}}(s) - s$ and $\mathcal{E}^{\mathrm{u}}(s)$ are $2\pi\varepsilon$-periodic with respect to s. Hence $\mathcal{S}^{\mathrm{u}}$ and $\mathcal{E}^{\mathrm{u}}$ can be analytically extended for all $s \in \mathbb{C}$ such that $|\operatorname{Im} s| \leq a - \varepsilon$.*

b) *Moreover, for $s \in \mathbb{R}$, $S = \mathcal{S}^{\mathrm{u}}(s)$ is real analytic and invertible, and its inverse $s = s^{\mathrm{u}}(S)$ satisfies that $s^{\mathrm{u}}(S) - S$ is $O(\mu\varepsilon^{p+1})$ and $2\pi\varepsilon$-periodic in S.*

PROOF. We prove it for $\mathcal{S}^{\mathrm{u}}(s) - s$. By property (6.3) and since $\mathcal{S}^{\mathrm{u}}(s) - s$ does not depend on t, we have that

$$\begin{aligned}
\mathcal{S}^{\mathrm{u}}(s + 2\pi\varepsilon) - (s + 2\pi\varepsilon) &= \mathcal{S}(\gamma^{\mathrm{u}}_{\mu,\varepsilon}(t, s + 2\pi\varepsilon), t/\varepsilon) - (t - t_0) - (s + 2\pi\varepsilon) \\
&= \mathcal{S}(\gamma^{\mathrm{u}}_{\mu,\varepsilon}(t + 2\pi\varepsilon, s), (t + 2\pi\varepsilon)/\varepsilon) - [(t + 2\pi\varepsilon) - t_0] - s \\
&= \mathcal{S}^{\mathrm{u}}(s) - s.
\end{aligned}$$

Analogously, $\mathcal{E}^{\mathrm{u}}(s + 2\pi\varepsilon) = \mathcal{E}^{\mathrm{u}}(s)$.

Next we prove the second part of this lemma. We recall that, by Theorem 4.2:

$$\mathcal{S}(\gamma^{\mathrm{s}}_{\mu,\varepsilon}(t,s), t/\varepsilon) = t - t_0 + s + \mu\varepsilon^{p+1}\mathcal{X}(s)$$

and that, for $(x,y) \in U$

$$\mathcal{S}(x,y,\theta) = \mathcal{S}_0(x,y) + O(\mu\varepsilon^{p+1})$$

where $\mathcal{S}_0$ is a flow box coordinate when $\mu = 0$. Also, by Theorem 5.1,

$$\gamma^{\mathrm{u}}_{\mu,\varepsilon}(t,s) - \gamma_0(t+s) = O(\mu\varepsilon^{\nu})$$

for any $t \in \mathbb{R}$ and $s \in \mathbb{C}$ such that $T \leq t + \operatorname{Re} s \leq 2T$ and $|\operatorname{Im} s| \leq a - \varepsilon$. Then we obtain that

$$\begin{aligned}
\mathcal{S}^{\mathrm{u}}(s) - s &= \mathcal{S}(\gamma^{\mathrm{u}}_{\mu,\varepsilon}(t,s), t/\varepsilon) - (t - t_0) - s \\
&= \mathcal{S}_0(\gamma_0(t+s)) - (t - t_0) - s + O(\mu\varepsilon^{\nu}, \mu\varepsilon^{p+1}) \\
&= O(\mu\varepsilon^{\nu}). \qquad (6.5)
\end{aligned}$$

Since $\mathcal{S}^{\mathrm{u}}(s) - s$ is $2\pi\varepsilon$-periodic in s and analytic in the complex strip $|\operatorname{Im} s| \leq a - \varepsilon$, we expand it in Fourier series,

$$\mathcal{S}^{\mathrm{u}}(s) - s = \sum_{k \in \mathbb{Z}} \mathcal{S}^{\mathrm{u}}_k(\varepsilon) e^{iks/\varepsilon}$$

with

$$\mathcal{S}_k^{\mathrm{u}}(\varepsilon) = \frac{1}{2\pi\varepsilon}\int_0^{2\pi\varepsilon} (\mathcal{S}^{\mathrm{u}}(s) - s)e^{-iks/\varepsilon}\,ds.$$

Moreover, for $s \in \mathbb{R}$ we can write

$$\mathcal{S}_k^{\mathrm{u}}(\varepsilon) = \frac{1}{2\pi\varepsilon}\int_0^{2\pi\varepsilon} \left[\mathcal{S}^{\mathrm{u}}(s \pm i(a-\varepsilon)) - (s \pm i(a-\varepsilon))\right] e^{-ik(s\pm i(a-\varepsilon))/\varepsilon}\,ds.$$

Thus, using the estimate (6.5), for $k \neq 0$, we obtain

$$\begin{aligned}\mathcal{S}_k^{\mathrm{u}}(\varepsilon) &= \frac{e^{-|k|(a-\varepsilon)/\varepsilon}}{2\pi\varepsilon}\int_0^{2\pi\varepsilon} \left[\mathcal{S}^{\mathrm{u}}(s \pm i(a-\varepsilon)) - (s \pm i(a-\varepsilon))\right] e^{-iks/\varepsilon}\,ds.\\ &= O(\mu\varepsilon^{\nu})e^{-|k|a/\varepsilon}.\end{aligned}$$

where we consider the sign $+$ for $k < 0$ and the sign $-$ for $k > 0$. Summing the Fourier series and applying the above equality, we deduce

$$\begin{aligned}\mathcal{S}^{\mathrm{u}}(s) - s &= \mathcal{S}_0^{\mathrm{u}}(\varepsilon) + O(\mu\varepsilon^{\nu})e^{-a/\varepsilon}\\ \frac{d\mathcal{S}^{\mathrm{u}}}{ds}(s) - 1 &= O(\mu\varepsilon^{\nu})e^{-a/\varepsilon},\end{aligned}$$

(taking into account that $\frac{d\mathcal{S}^{\mathrm{u}}}{ds}(s) - 1$ has zero mean).

This implies that $S = \mathcal{S}^{\mathrm{u}}(s)$ is invertible. On the other hand, if $s \in \mathbb{R}$, by Remark 5.2 we have that

$$\gamma_{\mu,\varepsilon}^{\mathrm{u}}(t,s) - \gamma_0(t+s) = O(\mu\varepsilon^{p+1}),$$

hence

$$\mathcal{S}^{\mathrm{u}}(s) - s = O(\mu\varepsilon^{p+1}).$$

Therefore $\mathcal{S}_0^{\mathrm{u}}(\varepsilon) = O(\mu\varepsilon^{p+1})$. We denote $s = s^{\mathrm{u}}(S)$ its inverse which is analytic. Moreover $s^{\mathrm{u}}(S) - S = O(\mu\varepsilon^{p+1})$. To see that $s^{\mathrm{u}}(S) - S$ is $2\pi\varepsilon$-periodic, we observe that, by a), $\mathcal{S}^{\mathrm{u}}(s + 2\pi\varepsilon) = \mathcal{S}^{\mathrm{u}}(s) + 2\pi\varepsilon$, thus

$$\begin{aligned}s^{\mathrm{u}}(S + 2\pi\varepsilon) - (S + 2\pi\varepsilon) &= s^{\mathrm{u}}(\mathcal{S}^{\mathrm{u}}(s) + 2\pi\varepsilon) - (S + 2\pi\varepsilon)\\ &= s^{\mathrm{u}}(\mathcal{S}^{\mathrm{u}}(s + 2\pi\varepsilon)) - (S + 2\pi\varepsilon)\\ &= s + 2\pi\varepsilon - (S + 2\pi\varepsilon)\\ &= s^{\mathrm{u}}(S) - S\end{aligned}$$

as we wanted. □

Now we define the splitting function. From Theorem 4.2 it follows that the local stable manifold $\gamma_\mu^{\mathrm{s}}(t,s)$ (for (t,s) such that $|\operatorname{Im} s| \leq a - \varepsilon$ and $T \leq t + \operatorname{Re} s \leq 2T$) can be written in the (S, E) coordinates:

(6.6) $$(S,E) = (\mathcal{S}(\gamma_{\mu,\varepsilon}^{\mathrm{s}}(t,s), t/\varepsilon), \mathcal{E}(\gamma_{\mu,\varepsilon}^{\mathrm{s}}(t,s), t/\varepsilon)) = (t - t_0 + s + \mu\varepsilon^{p+1}\mathcal{X}(s), 0)$$

and the local unstable manifold $\gamma_{\mu,\varepsilon}^{\mathrm{u}}(t,s)$ (for (t,s) such that $|\operatorname{Im} s| \leq a - \varepsilon$ and $T \leq t + \operatorname{Re} s \leq 2T$) can be expressed as

$$(S,E) = (\mathcal{S}(\gamma_{\mu,\varepsilon}^{\mathrm{u}}(t,s), t/\varepsilon), \mathcal{E}(\gamma_{\mu,\varepsilon}^{\mathrm{u}}(t,s), t/\varepsilon)) = (t - t_0 + \mathcal{S}^{\mathrm{u}}(s), \mathcal{E}^{\mathrm{u}}(s)).$$

We consider the Poincaré map

$$P_{\mu,\varepsilon}^{t_0}(x,y) = \varphi_{\mu,\varepsilon}(2\pi\varepsilon + t_0, t_0, x, y),$$

where $\varphi_{\mu,\varepsilon}(t, t_0, x, y)$ is the solution of system (6.1).

The restriction to U of the unstable curve C^{u} of $P^{t_0}_{\mu,\varepsilon}$, is given by $\gamma^{\mathrm{u}}_{\mu,\varepsilon}(t_0,s)$ parameterized by $s \in \mathbb{C}$ such that $T \leq \operatorname{Re} s \leq 2T$ and $|\operatorname{Im} s| \leq a-\varepsilon$. Indeed, let $W^{*,+}(P^{t_0}_{\mu,\varepsilon},0)$ $(*=\mathrm{s},\mathrm{u})$ be the right hand side of the stable and the unstable invariant curves of the origin of the map $P^{t_0}_{\mu,\varepsilon}$. Since the parameterizations $\gamma^{\mathrm{s,u}}_{\mu,\varepsilon}$, as functions of t are solutions of system (6.1), we have that

$$\begin{aligned} C^{\mathrm{s}} &\equiv \{\gamma^{\mathrm{s}}_{\mu,\varepsilon}(t_0,s) : \operatorname{Re} s + t_0 \geq T, \quad |\operatorname{Im} s| \leq a-\varepsilon\} \subset W^{\mathrm{s},+}(P^{t_0}_{\mu,\varepsilon},0) \\ C^{\mathrm{u}} &\equiv \{\gamma^{\mathrm{u}}_{\mu,\varepsilon}(t_0,s) : \operatorname{Re} s + t_0 \leq -T, \quad |\operatorname{Im} s| \leq a-\varepsilon\} \subset W^{\mathrm{u},+}(P^{t_0}_{\mu,\varepsilon},0). \end{aligned}$$

Moreover, since

$$\gamma^{*}_{\mu,\varepsilon}(t+2\pi\varepsilon,s) = \gamma^{*}_{\mu,\varepsilon}(t,s+2\pi\varepsilon), \qquad * = \mathrm{s},\mathrm{u},$$

in their respective domains, we have that

$$P^{t_0}_{\mu,\varepsilon}(\gamma^{*}_{\mu,\varepsilon}(t_0,s)) = \gamma^{*}_{\mu,\varepsilon}(2\pi\varepsilon+t_0,s) = \gamma^{*}_{\mu,\varepsilon}(t_0,s+2\pi\varepsilon),$$

which means that if we consider s as the variable in $C^{\mathrm{s}} \subset W^{\mathrm{s},+}(P^{t_0}_{\mu,\varepsilon},0)$ the dynamics of $P^{t_0}_{\mu,\varepsilon}$ on C^{s} is just

$$s \longmapsto s+2\pi\varepsilon.$$

Therefore in the (S,E) variables, C^{u} is represented by

$$(S,E) = (\mathcal{S}(\gamma^{\mathrm{u}}_{\mu,\varepsilon}(t_0,s),t_0/\varepsilon), \mathcal{E}(\gamma^{\mathrm{u}}_{\mu,\varepsilon}(t_0,s),t_0/\varepsilon)) = (\mathcal{S}^{\mathrm{u}}(s),\mathcal{E}^{\mathrm{u}}(s)).$$

By property b) of Lemma 6.1, the equality $S = \mathcal{S}^{\mathrm{u}}(s)$ can be inverted for values of s such that $|\operatorname{Im} s| < a-\varepsilon$, $s = s^{\mathrm{u}}(S)$, thus the function ϕ which gives the variable E as a function of S, is, in fact, defined explicitly by:

$$\phi(S) = \mathcal{E}^{\mathrm{u}}(s^{\mathrm{u}}(S)). \tag{6.7}$$

We observe that the splitting function is $2\pi\varepsilon$-periodic and hence is defined on $\mathbb{R}$.

The parameterization of the unstable manifold, $\gamma^{\mathrm{u}}_{\mu,\varepsilon}(t,s)$, introduced in Theorem 3.1 is not uniquely determined. Indeed, if we take $s = S+\varrho(S)$ where ϱ is a $2\pi\varepsilon$-periodic function which is $O(\mu\varepsilon^{p+1})$, then $\bar{\gamma}^{\mathrm{u}}_{\mu,\varepsilon}(t,S) = \gamma^{\mathrm{u}}_{\mu,\varepsilon}(t,S+\varrho(S))$ is another parameterization which also satisfies all properties we have proved until now.

Since $s^{\mathrm{u}}(S)-S$ is $O(\mu\varepsilon^{p+1})$ and $2\pi\varepsilon$-periodic in S we can introduce the new parameterization for the unstable manifold

$$\tilde{\gamma}^{\mathrm{u}}_{\mu,\varepsilon}(t,S) = \gamma^{\mathrm{u}}_{\mu,\varepsilon}(t,s^{\mathrm{u}}(S)).$$

We do not change the parameterization for the stable manifold

$$\tilde{\gamma}^{\mathrm{s}}_{\mu,\varepsilon}(t,S) = \gamma^{\mathrm{s}}_{\mu,\varepsilon}(t,S).$$

Finally, after this change of parameter, the splitting function defined in (6.7) can also be represented in the form

$$\phi(S) = \mathcal{E}(\tilde{\gamma}^{\mathrm{u}}_{\mu,\varepsilon}(t,S),t/\varepsilon). \tag{6.8}$$

Now we state two technical lemmas from **[DS2]**. Given $(t,s) \in D^{\mathrm{ext}}_{\varepsilon}$ and $\xi : D^{\mathrm{ext}}_{\varepsilon} \to \mathbb{C}^2$, we introduce $\tau = |t+s-ia|$ and

$$|\xi(t,s)|_{\tau} = |\xi_1(t,s)| + \tau|\xi_2(t,s)|.$$

LEMMA 6.2. *For t, t_0, l real and s complex, such that $0 \leq \operatorname{Im} s < a$ and*

$$-2T \leq t_0 + \operatorname{Re} s \leq t + \operatorname{Re} s \leq 2T, \qquad t_0 + \operatorname{Re} s < 0$$

we denote

$$\rho_{[t_0,t]}^{-l}(s) \equiv \begin{cases} \sup \dfrac{1}{|\sigma + s - ia|^l}, & \text{if } l \neq 0 \\ \sup |\ln(|\sigma + s - ia|)|, & \text{if } l = 0 \end{cases}$$

where the supremum is taken for $\sigma \in [t_0, t]$.

Then there exists a constant K which only depends on l such that

$$\int_{t_0}^{t} \frac{d\sigma}{|\sigma + s - ia|^l} \leq K \rho_{[t_0,t]}^{-(l-1)}(s). \tag{6.9}$$

LEMMA 6.3. *Let $\delta_0 \in (0,1)$ and let $\delta : [0,+\infty) \to \mathbb{R}$ be a function such that $\delta(\tau) \leq \delta_0/\tau^{r-1}$. Suppose that $\xi(t,s)$ and $\bar{\xi}(t,s)$ are two functions defined in $D_\varepsilon^{\text{ext}}$. We will write $\xi(t,s) = \xi = (\xi_1, \xi_2)$ and $\bar{\xi}(t,s) = \bar{\xi} = (\bar{\xi}_1, \bar{\xi}_2)$. Assume that*

$$|\xi|_\tau, |\bar{\xi}|_\tau \leq \delta(\tau).$$

Then, we have that

$$\begin{aligned} |V'(\alpha_0(t+s) + \xi_1) - V'(\alpha_0(t+s) + \bar{\xi}_1)| &\leq K \frac{|\xi_1 - \bar{\xi}_1|}{\tau^2}, && (t,s) \in D_\varepsilon^{\text{ext}}, \\ |g(\gamma_0(t+s) + \xi, t/\varepsilon) - g(\gamma_0(t+s) + \bar{\xi}, t/\varepsilon)|_\tau &\leq K \frac{|\xi - \bar{\xi}|_\tau}{\tau^{\ell - 2r + 1}}, && (t,s) \in D_\varepsilon^{\text{ext}}, \end{aligned}$$

where $g(x,y,t/\varepsilon) = (\partial_y h_1(x,y,t/\varepsilon), -\partial_x h_1(x,y,t/\varepsilon))$.

REMARK 6.1. *Since*

$$-V'(\alpha_0(u)) = \dot{\beta}_0(u) = \ddot{\alpha}_0(u)$$

has a singularity of order $r+1$ at $u = ia$, we have that, for $(t,s) \in D_\varepsilon^{\text{ext}}$ such that $0 \leq \operatorname{Im} s < a$:

$$|V^{(j+1)}(\alpha_0(t+s))| \leq K \frac{1}{\tau^{2-(j-1)(r-1)}}, \qquad \text{for } j \geq 0. \tag{6.10}$$

*By hypothesis **HP3**, $h_1(x,y,\theta)$ is a polynomial in (x,y). When we evaluate h_1 at $(x,y) = \gamma_0(u)$, by the definition of ℓ in Chapter 1, the function has a singularity of order at most ℓ at $u = ia$, hence for (t,s) as before*

$$|\partial_x^{k_1} \partial_y^{k_2} h_1(\gamma_0(t+s), t/\varepsilon)| \leq K \frac{1}{\tau^{\ell - k_1(r-1) - k_2 r}}, \qquad \text{for } k_1,\ k_2 \geq 0. \tag{6.11}$$

Since $h_1(x,y,t/\varepsilon)$ is $2\pi\varepsilon$-periodic in t, the Melnikov function,

$$M(s,\varepsilon) = \int_{-\infty}^{+\infty} \{h_0, h_1\}(\gamma_0(t+s), t/\varepsilon)\, dt,$$

has the same periodicity with respect to s. We denote by $M_k(\varepsilon)$ its Fourier's coefficients, i.e.,

$$M(s,\varepsilon) = \sum_{k \in \mathbb{Z}} M_k(\varepsilon) e^{iks/\varepsilon}.$$

The next proposition asserts that the Melnikov function is a good approximation of $\mathcal{E}^{\mathrm{u}}(s)$ for $|\operatorname{Im} s| \leq a - \varepsilon$ and gives that, in particular when $s \in \mathbb{R}$, the approximation is exponentially small.

PROPOSITION 6.1. *Under hypotheses **HP1-HP5**, $\mathcal{S}^{\mathrm{u}}$ and $\mathcal{E}^{\mathrm{u}}$ satisfy the following estimates:*

a) For $s \in \mathbb{C}$ such that $|\operatorname{Im} s| \leq a - \varepsilon$,

$$\mathcal{E}^{\mathrm{u}}(s) = \mu\varepsilon^p M(s,\varepsilon) + O(\mu^2\varepsilon^{2\nu+r-1}, \mu\varepsilon^{p+1}).$$

b) Let $\mathcal{E}_0^{\mathrm{u}}(\varepsilon) = \frac{1}{2\pi\varepsilon}\int_0^{2\pi\varepsilon} \mathcal{E}^{\mathrm{u}}(s)\, ds$. For $s \in \mathbb{R}$,

$$\mathcal{E}^{\mathrm{u}}(s) - \mathcal{E}_0^{\mathrm{u}}(\varepsilon) = \mu\varepsilon^p M(s,\varepsilon) + O(\mu^2\varepsilon^{2\nu+r-1}, \mu\varepsilon^{p+1})e^{-a/\varepsilon}.$$

PROOF. In Theorem 4.2 we have proved that

$$\mathcal{E}(\gamma^{\mathrm{s}}_{\mu,\varepsilon}(t,s), t/\varepsilon) = 0 \tag{6.12}$$

and

$$\mathcal{E}(x,y,\theta) = h_0(x,y) + O(\mu\varepsilon^{p+1}). \tag{6.13}$$

Since $\mathcal{E}^{\mathrm{u}}(s)$ does not depend on t, for any s we choose $t = T_s$ with $T_s = T - \operatorname{Re} s$ and therefore, for $(t,s) = (T_s, s)$, $\gamma^{\mathrm{u}}_{\mu,\varepsilon}(t,s)$ and $\gamma^{\mathrm{s}}_{\mu,\varepsilon}(t,s)$ belong to the domain of the flow box coordinates U. Then, from the definition (6.4) of $\mathcal{E}^{\mathrm{u}}$ and properties (6.12) and (6.13):

$$\begin{aligned}\mathcal{E}^{\mathrm{u}}(s) &= \mathcal{E}(\gamma^{\mathrm{u}}_{\mu,\varepsilon}(t,s), t/\varepsilon) - \mathcal{E}(\gamma^{\mathrm{s}}_{\mu,\varepsilon}(t,s), t/\varepsilon) \\ &= h_0(\gamma^{\mathrm{u}}_{\mu,\varepsilon}(t,s)) - h_0(\gamma^{\mathrm{s}}_{\mu,\varepsilon}(t,s)) + O(\mu\varepsilon^{p+1}),\end{aligned} \tag{6.14}$$

if $|\operatorname{Im} s| \leq a - \varepsilon$.

Since, for any s such that $|\operatorname{Im} s| \leq a - \varepsilon$, we have that

$$\begin{array}{lcll}\gamma^{\mathrm{s}}_{\mu,\varepsilon}(t,s) & \to & 0 & \text{when } t \to +\infty \\ \gamma^{\mathrm{u}}_{\mu,\varepsilon}(t,s) & \to & 0 & \text{when } t \to -\infty\end{array}$$

we deduce

$$\lim_{t\to+\infty} h_0(\gamma^{\mathrm{s}}_{\mu,\varepsilon}(t,s)) = \lim_{t\to-\infty} h_0(\gamma^{\mathrm{u}}_{\mu,\varepsilon}(t,s)) = 0.$$

Then

$$\begin{aligned}&h_0(\gamma^{\mathrm{u}}_{\mu,\varepsilon}(T_s,s)) - h_0(\gamma^{\mathrm{s}}_{\mu,\varepsilon}(T_s,s)) \\ &\quad= \int_{-\infty}^{T_s} \partial_t \left[h_0(\gamma^{\mathrm{u}}_{\mu,\varepsilon}(t,s))\right] dt - \int_{T_s}^{+\infty} \partial_t \left[h_0(\gamma^{\mathrm{s}}_{\mu,\varepsilon}(t,s))\right] dt \\ &\quad= \mu\varepsilon^p \left[\int_{-\infty}^{T_s} \{h_0,h_1\}(\gamma^{\mathrm{u}}_{\mu,\varepsilon}(t,s), t/\varepsilon)\, dt + \int_{T_s}^{+\infty} \{h_0,h_1\}(\gamma^{\mathrm{s}}_{\mu,\varepsilon}(t,s), t/\varepsilon)\, dt\right].\end{aligned}$$

Adding and subtracting the Melnikov function we can write

$$
\begin{aligned}
h_0(\gamma^{\mathrm{u}}_{\mu,\varepsilon}(T_s,s)) &- h_0(\gamma^{\mathrm{s}}_{\mu,\varepsilon}(T_s,s)) \\
&= \mu\varepsilon^p \int_{-\infty}^{-T_s} \{h_0,h_1\}(\gamma^{\mathrm{u}}_{\mu,\varepsilon},t/\varepsilon) - \{h_0,h_1\}(\gamma_0,t/\varepsilon)\,dt \\
&+ \mu\varepsilon^p \int_{-T_s}^{T_s} \{h_0,h_1\}(\gamma^{\mathrm{u}}_{\mu,\varepsilon},t/\varepsilon) - \{h_0,h_1\}(\gamma_0,t/\varepsilon)\,dt \qquad (6.15)\\
&+ \mu\varepsilon^p \int_{T_s}^{+\infty} \{h_0,h_1\}(\gamma^{\mathrm{s}}_{\mu,\varepsilon},t/\varepsilon) - \{h_0,h_1\}(\gamma_0,t/\varepsilon)\,dt \\
&+ \mu\varepsilon^p \int_{-\infty}^{+\infty} \{h_0,h_1\}(\gamma_0,t/\varepsilon)\,dt
\end{aligned}
$$

where $\gamma^{\mathrm{u}}_{\mu,\varepsilon}$, $\gamma^{\mathrm{s}}_{\mu,\varepsilon}$ and γ_0 denote $\gamma^{\mathrm{u}}_{\mu,\varepsilon}(t,s)$, $\gamma^{\mathrm{s}}_{\mu,\varepsilon}(t,s)$ and $\gamma_0(t+s)$ respectively. By conclusion 3) of Theorem 3.1, the first and the third lines of the right hand side of (6.15) are $O(\mu^2\varepsilon^{2p+1})$. It remains to bound the second line.

It is not difficult to see that, if we write $\gamma^*_{\mu,\varepsilon}(t,s) = (\alpha^*_{\mu,\varepsilon}(t,s),\beta^*_{\mu,\varepsilon}(t,s))$ for $* = \mathrm{s},\mathrm{u}$,

$$
\begin{aligned}
&\{h_0,h_1\}(\gamma^*_{\mu,\varepsilon},t/\varepsilon) - \{h_0,h_1\}(\gamma_0,t/\varepsilon) \\
&= V'(\alpha^*_{\mu,\varepsilon})[\partial_y h_1(\gamma^*_{\mu,\varepsilon},t/\varepsilon) - \partial_y h_1(\gamma_0,t/\varepsilon)] + [V'(\alpha^*_{\mu,\varepsilon}) - V'(\alpha_0)]\partial_y h_1(\gamma_0,t/\varepsilon) \\
&- \beta^*_{\mu,\varepsilon}[\partial_x h_1(\gamma^*_{\mu,\varepsilon},t/\varepsilon) - \partial_x h_1(\gamma_0,t/\varepsilon)] - (\beta^*_{\mu,\varepsilon} - \beta_0)\partial_x h_1(\gamma_0,t/\varepsilon).
\end{aligned}
$$

Using bounds (6.10), (6.11) and Lemma 6.3 and taking into account that, by the extension theorem, $\gamma^{\mathrm{u}}_{\mu,\varepsilon} - \gamma_0 = O(\mu\varepsilon^\nu)$, we get

$$
|\{h_0,h_1\}(\gamma^{\mathrm{u}}_{\mu,\varepsilon},t/\varepsilon) - \{h_0,h_1\}(\gamma_0,t/\varepsilon)| \le K\frac{\mu\varepsilon^\nu}{\tau^{\ell-r+2}}
$$

(we recall that $\tau = |t+s-ia|$). Then, applying the estimate (6.9) with $l = \ell - r + 2$ we obtain that the second line in (6.15) is $O(\mu\varepsilon^{p+\nu-\ell+r-1})$. Thus

$$
h_0(\gamma^{\mathrm{u}}_{\mu,\varepsilon}(T_s,s)) - h_0(\gamma^{\mathrm{s}}_{\mu,\varepsilon}(T_s,s)) = \mu\varepsilon^p M(s,\varepsilon) + O(\mu^2\varepsilon^{2p+1},\mu^2\varepsilon^{2\nu+r-1}).
$$

Now a) follows from (6.14) and from the previous expression. Note that, since $\ell \ge r-1$, one has $2p+1 \ge 2\nu + r - 1$.

To prove b), since $\mathcal{E}^{\mathrm{u}}(s)$ is $2\pi\varepsilon$-periodic in s and analytic in the complex strip $|\operatorname{Im} s| \le a - \varepsilon$, we expand it in Fourier series

$$
\mathcal{E}^{\mathrm{u}}(s) = \sum_{k\in\mathbb{Z}} \mathcal{E}^{\mathrm{u}}_k(\varepsilon) e^{iks/\varepsilon}.
$$

It is clear that, for $s \in \mathbb{R}$ we can write

$$
\mathcal{E}^{\mathrm{u}}_k(\varepsilon) = \frac{1}{2\pi\varepsilon}\int_0^{2\pi\varepsilon} \mathcal{E}^{\mathrm{u}}(s \pm i(a-\varepsilon)) e^{-ik(s\pm i(a-\varepsilon))/\varepsilon}\,ds.
$$

Thus, by the conclusion a) of this proposition about the estimate of $\mathcal{E}^{\mathrm{u}}(s)$ in the complex domain, for $k \ne 0$ we obtain

$$
\begin{aligned}
\mathcal{E}^{\mathrm{u}}_k(\varepsilon) &= \frac{e^{-|k|(a-\varepsilon)/\varepsilon}}{2\pi\varepsilon}\int_0^{2\pi\varepsilon} \mathcal{E}^{\mathrm{u}}(s \pm i(a-\varepsilon)) e^{-iks/\varepsilon}\,ds \\
&= \mu\varepsilon^p M_k(\varepsilon) + O(\mu^2\varepsilon^{2\nu+r-1},\mu\varepsilon^{p+1}) e^{-|k|(a-\varepsilon)/\varepsilon},
\end{aligned}
$$

where we consider the sign $+$ for $k < 0$ and the sign $-$ for $k > 0$. Here $M_k(\varepsilon)$ are the Fourier coefficients of the Melnikov function. Now b) follows summing the Fourier series and applying the above equality. □

6.3. Proof of Theorem 1.1 and its corollary

First we will show that the function ϕ given in (6.7) can be used to measure some magnitudes related to the splitting. Then we will prove the formulas in Theorem 1.1. In the next proposition we prove the existence of primary homoclinic points and we relate the angle between the invariant manifolds and the area of the lobes with the splitting function. Moreover, we relate it with the Melnikov function. First we state a technical lemma which we will prove in Section 6.4.

LEMMA 6.4. *Under the standing conditions we have*

$$\mu\varepsilon^p M(S,\varepsilon) = \mu\varepsilon^\nu \sum_{k\in\mathbb{Z}\setminus\{0\}} e^{-|k|a/\varepsilon} M_k e^{iks/\varepsilon}$$

with $M_k = O(1)$ uniformly in k. Therefore,

$$\mu\varepsilon^p \frac{dM}{dS}(S,\varepsilon) = O(\mu\varepsilon^{\nu-1})e^{-a/\varepsilon}.$$

PROPOSITION 6.2. *The function $\phi : \mathbb{R} \to \mathbb{R}$ is $2\pi\varepsilon$-periodic, real analytic and satisfies the following properties:*

a) *There exists $h^{\mathrm{u}} \in \mathbb{R}$ such that $\gamma^{\mathrm{u}}_{\mu,\varepsilon}(t,h^{\mathrm{u}}) = \gamma^{\mathrm{s}}_{\mu,\varepsilon}(t,h^{\mathrm{s}})$, for all t (giving a homoclinic orbit), with $h^{\mathrm{s}} = \mathcal{S}^{\mathrm{u}}(h^{\mathrm{u}})$. For $n \in \mathbb{N}$, we define*

$$h^{\mathrm{s}}_n = h^{\mathrm{s}} + 2\pi\varepsilon n$$

which give homoclinic points. Clearly, for all n, $\phi(h^{\mathrm{s}}_n) = 0$. Moreover, $\phi'(h^{\mathrm{s}}_n)$ is independent of n, and

$$\begin{aligned}\phi'(h^{\mathrm{s}}_n) &= \partial_S\tilde{\gamma}^{\mathrm{s}}_{\mu,\varepsilon}(t,h^{\mathrm{s}}_n)\wedge\partial_S\tilde{\gamma}^{\mathrm{u}}_{\mu,\varepsilon}(t,h^{\mathrm{s}}_n)(1+O(\mu\varepsilon^{p+1}))\\ &= \|\partial_S\tilde{\gamma}^{\mathrm{s}}_{\mu,\varepsilon}(t,h^{\mathrm{s}}_n)\|\;\|\partial_S\tilde{\gamma}^{\mathrm{u}}_{\mu,\varepsilon}(t,h^{\mathrm{s}}_n)\|\sin\vartheta(t,h^{\mathrm{s}}_n)(1+O(\mu\varepsilon^{p+1})),\end{aligned}$$

for all t, where $\wedge$ denotes the exterior product on $\mathbb{R}^2$, and $\vartheta(t,h^{\mathrm{s}}_n)$ is the angle between $\partial_s\tilde{\gamma}^{\mathrm{u}}_{\mu,\varepsilon}(t,h^{\mathrm{s}}_n)$ and $\partial_s\tilde{\gamma}^{\mathrm{s}}_{\mu,\varepsilon}(t,h^{\mathrm{s}}_n)$.

b) *The area of the lobe between the invariant curves is given by*

$$A = \left|\int_h^{\bar h}\phi(S)\,dS\right|,$$

where h and $\bar h$ are two consecutive zeros of $\phi(S)$.

c) $\phi_0 = \int_{h_n}^{h_n+2\pi\varepsilon}\phi(S)\,dS = 0$.

d) *For $S \in \mathbb{R}$, $\phi(S)$ satisfies the estimate*

$$\phi(S) \equiv \mathcal{E}^{\mathrm{u}}(s^{\mathrm{u}}(S)) = \mu\varepsilon^p M(S,\varepsilon) + O(\mu^2\varepsilon^{2\nu+r-1},\mu^2\varepsilon^{\nu+p},\mu\varepsilon^{p+1})e^{-a/\varepsilon}.$$

REMARK 6.2. *In particular the splitting function ϕ is an instrument to study the transversality of the intersections.*

PROOF. We begin by proving the existence of homoclinic orbits. Let $P^{t_0}_{\mu,\varepsilon}$ be the Poincaré map

$$P^{t_0}_{\mu,\varepsilon}(x,y) = \varphi_{\mu,\varepsilon}(2\pi\varepsilon + t_0, t_0, x, y).$$

Since $P^{t_0}_{\mu,\varepsilon}$ is area preserving and $P^{t_0}_{0,\varepsilon}$ has a homoclinic connection (which coincides with the homoclinic orbit of the unperturbed differential equation), a well known geometric argument, applied to $P^{t_0}_{\mu,\varepsilon}$ restricted to the reals, gives that $P^{t_0}_{\mu,\varepsilon}$ has (real) primary homoclinic points. Since the iterates of the homoclinic points are also homoclinic points there will be such points in U.

Then there exist h^{u}, $h^{\mathrm{s}} \in \mathbb{R}$, $T \leq h^{\mathrm{u}} + t_0, h^{\mathrm{s}} + t_0 \leq 2T$, which depend on t_0, such that

$$z^h = \gamma^{\mathrm{u}}_{\mu,\varepsilon}(t_0, h^{\mathrm{u}}) = \gamma^{\mathrm{s}}_{\mu,\varepsilon}(t_0, h^{\mathrm{s}}).$$

Hence

$$\gamma^{\mathrm{u}}_{\mu,\varepsilon}(t, h^{\mathrm{u}}) = \gamma^{\mathrm{s}}_{\mu,\varepsilon}(t, h^{\mathrm{s}})$$

are defined for all $t \in \mathbb{R}$ and are a homoclinic solution of (6.1).

Given a homoclinic point z^h of $P^{t_0}_{\mu,\varepsilon}$ we can express it as $\gamma^{\mathrm{s}}_{\mu,\varepsilon}(t_0+2\pi\varepsilon, h^{\mathrm{s}}-2\pi\varepsilon)$. This implies that $h^{\mathrm{s}}(t_0) = h^{\mathrm{s}}(t_0 + 2\pi\varepsilon) + 2\pi\varepsilon$ and hence z^h is the homoclinic point of $P^{t_0+2\pi\varepsilon}_{\mu,\varepsilon}$ corresponding to $h^{\mathrm{s}}(t_0) - 2\pi\varepsilon$.

By Theorem 4.2, we can choose $s_0 = h^{\mathrm{s}}$. Therefore, taking t such that $T \leq t + h^{\mathrm{u}}, t + h^{\mathrm{s}} \leq 2T$, we can write

$$\mathcal{S}(\gamma^{\mathrm{s}}_{\mu,\varepsilon}(t, h^{\mathrm{s}}), t/\varepsilon) = t - t_0 + h^{\mathrm{s}},$$

and

$$h^{\mathrm{s}} = \mathcal{S}(\gamma^{\mathrm{s}}_{\mu,\varepsilon}(t, h^{\mathrm{s}}), t/\varepsilon) - (t - t_0) = \mathcal{S}(\gamma^{\mathrm{u}}_{\mu,\varepsilon}(t, h^{\mathrm{u}}), t/\varepsilon) - (t - t_0) = \mathcal{S}^{\mathrm{u}}(h^{\mathrm{u}}).$$

Moreover, by the definition of ϕ in (6.7) and the definition of $\mathcal{E}^{\mathrm{u}}$ in (6.4) we have

$$\begin{aligned} \phi(h^{\mathrm{s}}) &= \phi(\mathcal{S}^{\mathrm{u}}(h^{\mathrm{u}})) = \mathcal{E}^{\mathrm{u}}(s^{\mathrm{u}}(\mathcal{S}^{\mathrm{u}}(h^{\mathrm{u}}))) \\ &= \mathcal{E}^{\mathrm{u}}(h^{\mathrm{u}}) = \mathcal{E}(\gamma^{\mathrm{u}}_{\mu,\varepsilon}(t, h^{\mathrm{u}}), t/\varepsilon) \\ &= \mathcal{E}(\gamma^{\mathrm{s}}_{\mu,\varepsilon}(t, h^{\mathrm{s}}), t/\varepsilon) = 0. \end{aligned}$$

By the $2\pi\varepsilon$-periodicity of ϕ, $\phi(h^{\mathrm{s}}_n) = \phi(h^{\mathrm{s}} + 2\pi\varepsilon n) = \phi(h^{\mathrm{s}}) = 0$.

Obviously ϕ' is also $2\pi\varepsilon$-periodic, thus $\phi'(h^{\mathrm{s}}_n)$ does not depend on n. Now we compute $\phi'(h^{\mathrm{s}})$. We recall that

$$\gamma^{\mathrm{u}}_{\mu,\varepsilon}(t, h^{\mathrm{u}}) = \gamma^{\mathrm{u}}_{\mu,\varepsilon}(t, s^{\mathrm{u}}(h^{\mathrm{s}})) = \tilde{\gamma}^{\mathrm{u}}_{\mu,\varepsilon}(t, h^{\mathrm{s}}).$$

Differentiating in (6.8) we obtain

$$\phi'(h^{\mathrm{s}}) = \partial_x \mathcal{E}(\tilde{\gamma}^{\mathrm{u}}_{\mu,\varepsilon}(t, h^{\mathrm{s}}), t/\varepsilon)\partial_S \tilde{\alpha}^{\mathrm{u}}(t, h^{\mathrm{s}}) + \partial_y \mathcal{E}(\tilde{\gamma}^{\mathrm{u}}_{\mu,\varepsilon}(t, h^{\mathrm{s}}), t/\varepsilon)\partial_S \tilde{\beta}^{\mathrm{u}}(t, h^{\mathrm{s}}) \tag{6.16}$$

where $\tilde{\gamma}^{\mathrm{u}}_{\mu,\varepsilon}(t, s) = (\tilde{\alpha}^{\mathrm{u}}(t, s), \tilde{\beta}^{\mathrm{u}}(t, s))$. Moreover, differentiating with respect to s in (6.6) we obtain

$$\begin{aligned} 1 + O(\mu\varepsilon^p) &= \partial_x \mathcal{S}(\gamma^{\mathrm{s}}_{\mu,\varepsilon}(t, S), t/\varepsilon)\partial_S \alpha^{\mathrm{s}}(t, S) + \partial_y \mathcal{S}(\gamma^{\mathrm{s}}_{\mu,\varepsilon}(t, S), t/\varepsilon)\partial_S \beta^{\mathrm{s}}(t, S) \\ 0 &= \partial_x \mathcal{E}(\gamma^{\mathrm{s}}_{\mu,\varepsilon}(t, S), t/\varepsilon)\partial_S \alpha^{\mathrm{s}}(t, S) + \partial_y \mathcal{E}(\gamma^{\mathrm{s}}_{\mu,\varepsilon}(t, S), t/\varepsilon)\partial_S \beta^{\mathrm{s}}(t, S) \end{aligned}$$

and from this, taking into account that the change $(x, y) \mapsto (S, E)$ is canonical, we get, when $S = h^{\mathrm{s}}$

$$\begin{aligned} \partial_S \alpha^{\mathrm{s}}(t, h^{\mathrm{s}})(1 + O(\mu\varepsilon^p)) &= \partial_y \mathcal{E}(\gamma^{\mathrm{s}}_{\mu,\varepsilon}(t, h^{\mathrm{s}}), t/\varepsilon) = \partial_y \mathcal{E}(\gamma^{\mathrm{u}}_{\mu,\varepsilon}(t, h^{\mathrm{u}}), t/\varepsilon) \\ &= \partial_y \mathcal{E}(\tilde{\gamma}^{\mathrm{u}}_{\mu,\varepsilon}(t, h^{\mathrm{s}}), t/\varepsilon) \\ \partial_S \beta^{\mathrm{s}}(t, h^{\mathrm{s}})(1 + O(\mu\varepsilon^p)) &= -\partial_x \mathcal{E}(\gamma^{\mathrm{s}}_{\mu,\varepsilon}(t, h^{\mathrm{s}}), t/\varepsilon) = -\partial_x \mathcal{E}(\tilde{\gamma}^{\mathrm{u}}_{\mu,\varepsilon}(t, h^{\mathrm{s}}), t/\varepsilon). \end{aligned}$$

Substituting the derivatives of $\mathcal{E}$ in (6.16) we obtain the formula stated in a).

In order to prove b) we recall that the change $\tilde{\mathcal{C}}$, given in Theorem 4.2, which transforms the initial coordinates (x, y) to the flow box coordinates (S, E), is canonical. Therefore.

$$A = \left| \int \int_{Lobe} dx\, dy \right| = \left| \int \int_{\tilde{\mathcal{C}}(Lobe)} dS\, dE \right|.$$

Moreover, since the Poincaré map $P^{t_0}_{\mu,\varepsilon}$ is orientation preserving, there exists at least one primary homoclinic point of $P^{t_0}_{\mu,\varepsilon}$ between $z^h = \gamma^{\rm s}_{\mu,\varepsilon}(t_0, h^{\rm s})$ and $P^{t_0}_{\mu,\varepsilon}(z^h)$. We denote this homoclinic point by $\gamma^{\rm s}_{\mu,\varepsilon}(t_0, \bar{h}^{\rm s})$. By definition of the splitting function, the area of a lobe, in (S, E) coordinates, is the area of the splitting function between two consecutive zeros of ϕ, hence

$$A = \left| \int_{h^{\rm s}}^{\bar{h}^{\rm s}} \phi(S)\, dS \right|$$

with $h^{\rm s}$ and $\bar{h}^{\rm s}$ are two consecutive zeros of ϕ.

The conclusion c) asserts that the splitting function has zero mean. To prove it we note that since $P^{t_0}_{\mu,\varepsilon}$ is area preserving, a standard geometric argument gives that the area of two consecutive lobes one inner and the other outer, coincide. Therefore c) follows from b) and the fact that the change $\tilde{\mathcal{C}}$ is canonical.

Now we prove estimate d). By estimate b) of Proposition 6.1 as well as by the definition of $\phi(S) = \mathcal{E}^{\rm u}(s^{\rm u}(S))$, we have that for real values of S

$$\phi(S) \equiv \mathcal{E}^{\rm u}(s^{\rm u}(S)) = \mathcal{E}^{\rm u}_0(\varepsilon) + \mu\varepsilon^p M(s^{\rm u}(S), \varepsilon) + O(\mu^2\varepsilon^{2\nu+r-1}, \mu\varepsilon^{p+1})e^{-a/\varepsilon},$$

where $\mathcal{E}^{\rm u}_0(\varepsilon)$ is the 0-Fourier coefficient of $\mathcal{E}^{\rm u}$. Thus, in order to prove Theorem 1.1 we need to estimate $\mathcal{E}^{\rm u}_0(\varepsilon)$. By Taylor's Theorem and Lemma 6.4 we have for $S \in \mathbb{R}$,

$$\begin{aligned} M(s^{\rm u}(S), \varepsilon) &= M(S + O(\mu\varepsilon^{p+1}), \varepsilon) \\ &= M(S, \varepsilon) + \int_0^1 \frac{dM}{dS}(S + \zeta O(\mu\varepsilon^{p+1}), \varepsilon) O(\mu\varepsilon^{p+1})\, d\zeta \\ &= M(S, \varepsilon) + O(\mu\varepsilon^{\nu})e^{-a/\varepsilon}. \end{aligned}$$

Therefore, we have that

$$\phi(S) = \mathcal{E}^{\rm u}_0(\varepsilon) + \mu\varepsilon^p M(S, \varepsilon) + O(\mu^2\varepsilon^{2\nu+r-1}, \mu^2\varepsilon^{\nu+p}, \mu\varepsilon^{p+1})e^{-a/\varepsilon}.$$

Finally since by c), $\phi_0 = 0$, and the average of M is zero we get

$$\mathcal{E}^{\rm u}_0(\varepsilon) = O(\mu^2\varepsilon^{2\nu+r-1}, \mu^2\varepsilon^{\nu+p}, \mu\varepsilon^{p+1})e^{-a/\varepsilon}$$

and the estimate d) is proved. □

The proof of Theorem 1.1 is an immediate consequence of Proposition 6.2.

6.4. Proof of Lemma 6.4

The proof of this lemma has significative differences from the proof of the corresponding lemma in [**DS2**]. As we pointed out before, we are considering the case such that the parameterization of the homoclinic orbit has a singularity which is a branching point. The proof of Lemma 6.4 is the place where this hypothesis has to be taken into account. Since $u = \pm ia$ are branching points, the homoclinic orbit is defined in a neighborhood of the singularities except a segment starting at them, and therefore we can not use the residue theory in order to estimate the Melnikov integral.

The case that the singularity is a pole also follows from this proof taking $q = 1$ below.

We recall the definition of J given in Chapter 1:

$$J(x, y, t/\varepsilon) \equiv \{h_0, h_1\}(x, y, t/\varepsilon) \sim \sum_{n\neq 0} J_n(x, y)e^{int/\varepsilon}$$

and that $J(\gamma_0(t + s), t/\varepsilon)$ has a singularity of order at most $\ell + 1$. We also observe that the perturbation $h_1(x, y, \theta)$ can be written as

$$h_1(x, y, \theta) = \sum_{k\leq|l|\leq\kappa, l\in\mathbb{N}^2} a_l(\theta)x^{l_1}y^{l_2}.$$

We recall that, by its definition, ℓ can be expressed in the form

$$\ell = j_1(r - 1) + j_2 r = j_2 + (j_1 + j_2)\frac{c}{q}, \tag{6.17}$$

where j_1 and j_2 are such that

$$j_1(r - 1) + j_2 r = \max\{l_1(r - 1) + l_2 r \ : \ l_1 + l_2 \geq k \ , \ l = (l_1, l_2) \ , \ a_l(\theta) \neq 0\}.$$

Now we write the Fourier's coefficients of $M(s, \varepsilon)$, $M_n(\varepsilon)$, in terms of the Fourier's coefficients of J evaluated at $\gamma_0(u)$: $J_n(\gamma_0(u))$. We note that, since J is only continuous with respect to t/ε, the Fourier's series may not converge, although their Fourier's coefficients are well defined. However, since $M(., \varepsilon)$ is analytic and $2\pi\varepsilon$-periodic with respect to s, its Fourier series does converge.

We claim that

$$M_n(\varepsilon) = \frac{1}{2\pi\varepsilon}\int_{-\infty}^{+\infty} e^{-iun/\varepsilon} J_{-n}(\gamma_0(u))\, du.$$

Indeed, by definition of $M(s, \varepsilon)$, we obtain

$$\begin{aligned}
M_n(\varepsilon) &= \frac{1}{2\pi\varepsilon}\int_0^{2\pi\varepsilon} \left(\int_{-\infty}^{+\infty} J(\gamma_0(t + s), t/\varepsilon)\, dt\right) e^{-ins/\varepsilon}\, ds \\
&= \frac{1}{2\pi\varepsilon}\int_0^{2\pi\varepsilon}\int_{-\infty}^{+\infty} e^{-ins/\varepsilon} J\Big(\gamma_0(u), \frac{u - s}{\varepsilon}\Big)\, du\, ds \\
&= \frac{1}{2\pi\varepsilon}\int_{-\infty}^{+\infty} e^{-iun/\varepsilon}\int_0^{2\pi\varepsilon} e^{in(u-s)/\varepsilon} J\Big(\gamma_0(u), \frac{u - s}{\varepsilon}\Big)\, ds\, du \\
&= \int_{-\infty}^{+\infty} e^{-iun/\varepsilon} J_{-n}(\gamma_0(u))\, du.
\end{aligned}$$

Since h_1 is a polynomial in x, y, near the singularities $u = \pm ia$, from the expression of γ_0 given in **HP1**, $J_n(\gamma_0(u))$ has the form:

$$\begin{aligned}
J_n(\gamma_0(u)) &= \frac{1}{(u \pm ia)^{\ell+1}}\Big(J^{\pm}_{n,0} + \sum_{m\geq 0} J^{\pm}_{n,m}(u \pm ia)^{\frac{m}{q}}\Big) \\
&= \sum_{-\infty<m\leq(j_1+j_2)c} \frac{J^{\pm}_{n,(j_1+j_2)c-m}}{(u \pm ia)^{\frac{m}{q}+j_2+1}}
\end{aligned} \tag{6.18}$$

where j_1 and j_2 are defined in (6.17) and $J^{\pm}_{n,(j_1+j_2)c-m}$ are coefficients which depend on ε and μ.

Now we proceed to evaluate the integrals

$$\int_{-\infty}^{+\infty} e^{-iun/\varepsilon} J_{-n}(\gamma_0(u))\, du. \tag{6.19}$$

We consider first the case $n < 0$. We choose the path of integration $\Gamma = \Gamma_1 \vee \Gamma_2 \vee \ldots \vee \Gamma_8$ as indicated in the figure:

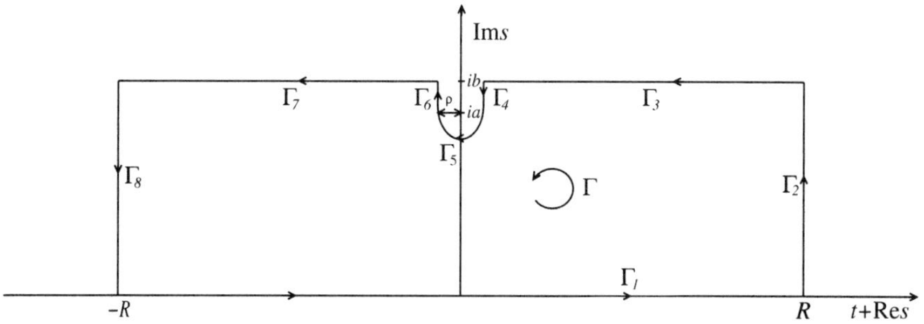

where $b > a$, ρ is small (obviously $\rho < a$) and R is big. Since we will play with the dependence of Γ on ρ, we will denote the path by $\Gamma(\rho)$.

Since the function $J_{-n}(\gamma_0(u))e^{-inu/\varepsilon}$ is analytic in the region enclosed by $\Gamma(\rho)$,

$$\int_{\Gamma(\rho)} J_{-n}(\gamma_0(u))e^{-inu/\varepsilon}\, du = 0, \qquad \forall \rho \in (0, \rho_0)$$

with ρ_0 small enough. The advantage of considering these curves is that the above integral does not depend on ρ. Therefore, in order to compute the dominant term of

$$\int_{\Gamma_1(\rho)} J_{-n}(\gamma_0(u))e^{-inu/\varepsilon}\, du = -\sum_{j=2}^{8} \int_{\Gamma_j(\rho)} J_{-n}(\gamma_0(u))e^{-inu/\varepsilon}\, du$$

the strategy consists on expanding the right hand side in terms of powers of ρ and then taking limit when ρ goes to zero. The terms with negative powers of ρ must cancel and the terms with positive powers of ρ tend to zero. Therefore we only have to take into account the coefficients of ρ^0 in such expansion.

We begin to look for the asymptotic expression of (6.19). First we observe that the integrals over Γ_2 and Γ_8 tend to zero when R tends to $+\infty$ and that the integrals over $\Gamma_3(\rho)$ and $\Gamma_7(\rho)$ are of order $O(e^{nb/\varepsilon})$, uniformly with respect to ρ.

Next we will compute the integrals over the paths $\Gamma_5(\rho)$, $\Gamma_4(\rho)$ and $\Gamma_6(\rho)$. For these three integrals we stay near the singularity ia, thus we can use the expansion of $J_{-n}(\gamma_0(u))$ given in (6.18). For $j = 4, 5, 6$, we have that

$$\int_{\Gamma_j(\rho)} J_{-n}(\gamma_0(u))e^{-inu/\varepsilon}\, du$$

$$= \sum_{m \le (j_1+j_2)c} J^{-}_{n,(j_1+j_2)c-m} \int_{\Gamma_j(\rho)} \frac{e^{-inu/\varepsilon}}{(u-ia)^{\frac{m}{q}+j_2+1}}\, du. \tag{6.20}$$

To evaluate the integrals in the right hand side of (6.20) we distinguish two cases: $m/q \notin \mathbb{N}$ and $m/q \in \mathbb{N}$. First we deal with the case $m/q \notin \mathbb{N}$:

1. Integral over $\Gamma_5(\rho)$. We parameterize this path by $g_5(\theta) = ia + \rho e^{-i\theta}$ with $\theta \in [0,\pi]$. Using the series expansion of the exponential we have that

$$\begin{aligned}\int_{\Gamma_5(\rho)} \frac{e^{-inu/\varepsilon}}{(u-ia)^{\frac{m}{q}+j_2+1}}\, du &= -ie^{na/\varepsilon}\rho \int_0^\pi \frac{e^{-in\rho e^{-i\theta}/\varepsilon} e^{-i\theta}}{(\rho e^{-i\theta})^{\frac{m}{q}+j_2+1}}\, d\theta \\ &= -ie^{na/\varepsilon}\rho^{-\frac{m}{q}-j_2} \sum_{l\geq 0} \int_0^\pi \frac{1}{l!}\left(\frac{-in\rho}{\varepsilon}\right)^l e^{i\theta(\frac{m}{q}+j_2-l)}\, d\theta.\end{aligned}$$

Therefore, since $m/q \notin \mathbb{N}$, the integral over $\Gamma_5(\rho)$ has no constant term in ρ, then this integral has no contribution to the ρ^0 term of its expansion.

2. Integral over $\Gamma_4(\rho)$. We parameterize this path by $g_4(\theta) = \rho - i\theta$ with $\theta \in [-b,-a]$. Then

$$\int_{\Gamma_4(\rho)} \frac{e^{-inu/\varepsilon}}{(u-ia)^{\frac{m}{q}+j_2+1}}\, du = -i\int_a^b \frac{e^{-\rho n i/\varepsilon} e^{n\theta/\varepsilon}}{(\rho + (\theta-a)i)^{\frac{m}{q}+j_2+1}}\, d\theta.$$

Given $l \in \mathbb{Z}^+$ and $\eta \in (0,1)$, we introduce the notation

$$I_l(\rho) = \int_a^b \frac{e^{n\theta/\varepsilon}}{(\rho + (\theta-a)i)^{\eta+l}}\, d\theta$$

and

$$f_l(\rho) = \frac{e^{na/\varepsilon}}{\rho^{\eta+l}} - \frac{e^{nb/\varepsilon}}{(\rho + i(b-a))^{\eta+l}}.$$

Integrating by parts in $I_l(\rho)$ we obtain a recurrence formula for $I_l(\rho)$:

$$I_l(\rho) = \frac{1}{i(\eta+l-1)}\left(f_{l-1}(\rho) + \frac{n}{\varepsilon} I_{l-1}\right)$$

and therefore

$$\begin{aligned}I_l(\rho) &= \sum_{j=1}^{l} \left(\frac{n}{\varepsilon}\right)^{j-1} \frac{1}{i^j} \frac{1}{(\eta+l-1)\cdots(\eta+l-j)} f_{l-j}(\rho) \\ &\quad + \left(\frac{n}{\varepsilon}\right)^l \frac{1}{i^l}\frac{1}{(\eta+l-1)\cdots\eta} I_0(\rho).\end{aligned} \tag{6.21}$$

The contribution of the j-term in the sum (6.21) to the ρ^0 term is

$$-\left(\frac{n}{\varepsilon}\right)^{j-1} \frac{1}{i^j}\frac{1}{(\eta+l-1)\cdots(\eta+l-j)} \frac{e^{nb/\varepsilon}}{(i(b-a))^{\eta+l-1}}. \tag{6.22}$$

Now we analyze $I_0(\rho)$. Note that, since $\rho > 0$, we have that $\arg(\rho + i(z-a)) \to \pi/2$ when $\rho \to 0$ for $z > a$. Therefore, using the dominated convergence theorem

$$I_0(\rho) = \int_a^b \frac{e^{n\theta/\varepsilon}}{(\rho + i(\theta-a))^\eta}\, d\theta \to e^{-\eta i\pi/2}\int_a^b \frac{e^{n\theta/\varepsilon}}{(\theta-a)^\eta}\, d\theta \qquad \text{when } \rho\to 0.$$

With elementary changes of variables we get

$$\int_a^b \frac{e^{n\theta/\varepsilon}}{(\theta-a)^\eta}\, d\theta = \left(\frac{\varepsilon}{|n|}\right)^{1-\eta} e^{na/\varepsilon}\int_0^{|n|(b-a)/\varepsilon} s^{-\eta}e^{-s}\, ds. \tag{6.23}$$

Moreover we have that

$$\int_0^{|n|(b-a)/\varepsilon} s^{-\eta}e^{-s}\, ds = \Gamma(1-\eta) - \psi(\varepsilon) \tag{6.24}$$

where Γ is the Gamma function and

$$\psi(\varepsilon) = \int_{|n|(b-a)/\varepsilon}^{+\infty} s^{-\eta} e^{-s}\, ds \le \left(\frac{\varepsilon}{|n|(b-a)}\right)^{\eta} e^{-|n|(b-a)/\varepsilon}$$

which is exponentially small. Using (6.22), (6.23) and (6.24), we obtain that the constant term in ρ of $I_l(\rho)$ is

$$\begin{aligned} I_l^0 &\equiv \sum_{j=1}^{l} \left(\frac{n}{\varepsilon}\right)^{j-1} \frac{1}{i^j} \frac{-e^{nb/\varepsilon}}{(\eta+l-1)\cdots(\eta+l-j)(i(b-a))^{\eta+j-1}} \\ &\quad + \left(\frac{n}{\varepsilon}\right)^{l} \frac{1}{i^l} \frac{1}{(\eta+l-1)\cdots\eta} \left(\frac{\varepsilon}{|n|}\right)^{1-\eta} e^{na/\varepsilon} e^{-\eta i\pi/2} [\Gamma(1-\eta) - \psi(\varepsilon)] \end{aligned}$$

and therefore

$$I_l^0 = \frac{(-1)^l}{i^l} \left(\frac{|n|}{\varepsilon}\right)^{l-1+\eta} e^{-\eta i\pi/2} \frac{\pi}{\sin(\pi\eta)} \frac{1}{\Gamma(l+\eta)} e^{-|n|a/\varepsilon}(1+O(\varepsilon))$$

where we have used the formula $\Gamma(1-\eta)\Gamma(\eta) = \pi/\sin(\eta\pi)$. Thus the constant term in ρ of the integral over $\Gamma_4(\rho)$ can be calculated with the previous estimates taking $l = j_2 + 1 + [m/q]$ and $\eta = m/q - [m/q]$, where $[\cdot]$ denotes the integer part function, and it becomes

$$-i\frac{(-1)^l}{i^l} \left(\frac{|n|}{\varepsilon}\right)^{l-1+\eta} e^{-\eta i\pi/2} \frac{\pi}{\sin(\pi\eta)} \frac{1}{\Gamma(l+\eta)} e^{-|n|a/\varepsilon}(1+O(\varepsilon)), \tag{6.25}$$

where the $O(\varepsilon)$ term is uniform with respect to n.

3. Integral over Γ_6. We parameterize this path by $g_6(\theta) = -\rho + i\theta$ with $\theta \in [a,b]$. Therefore,

$$\int_{\Gamma_6(\rho)} \frac{e^{-inu/\varepsilon}}{(u-ia)^{\frac{m}{q}+j_2+1}}\, du = i\int_a^b \frac{e^{\rho ni/\varepsilon} e^{n\theta/\varepsilon}}{(-\rho+(\theta-a)i)^{\frac{m}{q}+j_2+1}}\, d\theta.$$

Thus, if we define

$$K_l(\rho) = \int_a^b \frac{e^{\rho ni/\varepsilon}}{(-\rho+(\theta-a)i)^{l+\eta}}\, d\theta$$

we have that $K_l(\rho) = I_l(-\rho)$, and, for $\eta > 0$, by using the previous computations we obtain

$$\begin{aligned} K_l(\rho) &= \sum_{j=1}^{l} \left(\frac{n}{\varepsilon}\right)^{j-1} \frac{1}{i^j} \frac{1}{(\eta+l-1)\cdots(\eta+l-j)} f_{l-j}(-\rho) \\ &\quad + \left(\frac{n}{\varepsilon}\right)^{l} \frac{1}{i^l} \frac{1}{(\eta+l-1)\cdots\eta} K_0. \end{aligned}$$

As before we calculate the constant term K_0 of $K_0(\rho)$. In this case, the argument of $-\rho + (z-a)i$ belongs to $(-3\pi/2, -\pi)$ and therefore,

$$K_0(\rho) = \int_a^b \frac{e^{n\theta/\varepsilon}}{(-\rho+i(\theta-a))^{\eta}}\, d\theta \to e^{\eta i 3\pi/2} \int_a^b \frac{e^{n\theta/\varepsilon}}{(\theta-a)^{\eta}}\, d\theta, \qquad \text{when } \rho \to 0.$$

Consequently, the constant term in ρ of the integral over $\Gamma_6(\rho)$ is

$$i\frac{(-1)^l}{i^l}\left(\frac{|n|}{\varepsilon}\right)^{l-1+\eta} e^{\eta i3\pi/2}\frac{\pi}{\sin(\pi\eta)}\frac{1}{\Gamma(l+\eta)}e^{-|n|a/\varepsilon}(1+O(\varepsilon)), \tag{6.26}$$

where the $O(\varepsilon)$ term is uniform with respect to n.

Now we consider the case such that $m/q \in \mathbb{N}$. Taking into account that, $m/q + j_2 + 1 \in \mathbb{N}$, the functions

$$\frac{e^{-inu/\varepsilon}}{(u-ia)^{\frac{m}{q}+j_2+1}}$$

have a pole of order $m/q + j_2 + 1$. Therefore we can apply the residue theory. In this case the integral over Γ_1 turns out to be

$$2\pi i\frac{1}{(m/q+j_2)!}\left(\frac{-in}{\varepsilon}\right)^{m/q+j_2} e^{-|n|a/\varepsilon}(1+O(\varepsilon)). \tag{6.27}$$

Now we compute the dominant term of $M_n(\varepsilon)$ for $n<0$. Let $\eta_\ell = \ell - [\ell]$. If $\ell \notin \mathbb{N}$ and $n<0$, by (6.25) and (6.26) we obtain that

$$\begin{aligned} M_n(\varepsilon) &= \sum_{m\le (j_1+j_2)c} J^-_{-n,(j_1+j_2)c-m}\int_{-\infty}^{\infty}\frac{e^{-inu/\varepsilon}}{(u-ia)^{m/q+j_2+1}}\,du \\ &= i^{[\ell]+1}\left(\frac{|n|}{\varepsilon}\right)^{\ell}\frac{2\pi}{\Gamma(\ell+1)}J^-_{-n,0}e^{\eta_\ell i\pi/2}e^{-|n|a/\varepsilon}(1+O(\varepsilon)). \end{aligned} \tag{6.28}$$

Note that $i^{[\ell]}e^{\eta_\ell i\pi/2} = i^\ell$. If $\ell \in \mathbb{N}$, using expression (6.27),

$$M_n(\varepsilon) = i^{\ell+1}\left(\frac{|n|}{\varepsilon}\right)^{\ell}\frac{2\pi}{\ell!}J^-_{-n,0}e^{-|n|a/\varepsilon}(1+O(\varepsilon)). \tag{6.29}$$

REMARK 6.3. *The expressions of $M_n(\varepsilon)$ in (6.28) and (6.29) are computed independently. We note that (6.28) goes to (6.29) when η_ℓ goes to zero. Therefore we can use (6.28) for all ℓ.*

For the case $n>0$, it is sufficient to observe that

$$M_n(\varepsilon) = \overline{M_{-n}(\varepsilon)} \quad \text{and} \quad J^-_{-n,0} = \overline{J^+_{n,0}}. \tag{6.30}$$

The result follows summing the Fourier series, taking into account that the terms $O(\varepsilon)$ are uniform in n.

6.5. Proof of Corollary 1.1

Assume the same hypotheses as Theorem 1.1 and the further hypothesis **HP6**. Then, $J^+_{1,0} = \overline{J^-_{-1,0}}$ are different from zero. We write $J^-_{1,0} = |J^-_{1,0}|e^{i\theta}$.

From (6.28) and (6.30) we have

$$\begin{aligned} &M(s,\varepsilon) \\ &= \sum_{n>0}\frac{2\pi}{\Gamma(\ell+1)}\left(\frac{n}{\varepsilon}\right)^{\ell} e^{-na/\varepsilon}\left[(-i)^{\ell+1}J^+_{-n,0}e^{ins/\varepsilon} + i^{\ell+1}J^-_{n,0}e^{-ins/\varepsilon}\right](1+O(\varepsilon)) \\ &= \varepsilon^{-\ell}e^{-a/\varepsilon}\frac{2\pi}{\Gamma(\ell+1)}2\,\mathrm{Re}(i^{\ell+1}J^-_{1,0}e^{-is/\varepsilon}) + O(\varepsilon^{-\ell+1}e^{-a/\varepsilon}). \end{aligned}$$

Since $i^{\ell+1}J^-_{1,0}e^{-is/\varepsilon} = |J^-_{1,0}|e^{i(\theta+(\ell+1)\pi/2)}e^{-is/\varepsilon}$ the formula for $M(s,\varepsilon)$ follows.

From Theorem 1.1, to compute the area we have to estimate

$$\begin{aligned}\int_{s_0}^{\bar{s}_0} M(s,\varepsilon)\,ds &= \varepsilon^{-\ell} e^{-a/\varepsilon}\left[\frac{4\pi}{\Gamma(\ell+1)}|J_{1,0}^-|\int_{s_0}^{\bar{s}_0}\cos(\theta+(\ell+1)\pi/2-s/\varepsilon)\,ds+O(\varepsilon)\right]\\ &= \varepsilon^{-\ell+1} e^{-a/\varepsilon}\left[\frac{8\pi}{\Gamma(\ell+1)}|J_{1,0}^-|+O(\varepsilon)\right].\end{aligned}$$

Summing the Fourier series corresponding to $M'(s,\varepsilon)$, in the same way as for $M(s,\varepsilon)$ we have

$$|M'(s,\varepsilon)| = \varepsilon^{-\ell-1}\frac{4\pi}{\Gamma(\ell+1)}|J_{1,0}^-|e^{-a/\varepsilon}\sin(\theta+(\ell+1)\pi/2-s/\varepsilon)+O(\varepsilon^{-\ell}e^{-a/\varepsilon}).$$

If $M(s_0,\varepsilon)=0$, then s_0 is such that $s_0/\varepsilon=\theta+(\ell+1)\pi/2+\pi/2+k\pi+O(\varepsilon)$, $k\in\mathbb{Z}$, and therefore, $\sin(\theta+(\ell+1)\pi/2-s/\varepsilon)=\pm1+O(\varepsilon)$. Using the formula for the angle in Theorem 1.1 we get the result.

References

[An] S. Angenent, *A variational interpretation of Melnikov's function and exponentially small separatrix splitting*, Symplectic geometry, London Math. Soc. Lecture Note Ser., vol 192, Cambridge Univ. Press, Cambridge 1993, pp. 5–35.

[Ar1] V.I. Arnold, *Instability of dynamical systems with several degrees of freedom*, Soviet Math. Dokl. **5** (1964), 581–585.

[AKN] V.I. Arnold, V.V. Koslov and A.I. Neishtadt, *Dynamical systems III*, Encyclopaedia of Mathematical Sciences, vol. 3, Springer, Berlín, 1988.

[CFN] J. Casasayas, E. Fontich and A. Nunes, *Homoclinic orbits to parabolic points*, NoDEA Nonlinear Differential Equations Appl. **4** (1997), 201–216.

[Ch] V. Chernov, *On separatrix splitting of some quadratic area-preserving maps of the plane*, Regular & Chaotic Dynamics **3** (**1**) (1998), 49–65.

[DG] A. Delshams and P. Gutiérrez, *Splitting potential and the Poincaré-Melnikov method for whiskered tori in Hamiltonian systems*, Nonlinear Science **10** (2000), 433–476.

[DGJS1] A. Delshams, V.G. Gelfreich, A. Jorba and T.M. Seara, *Exponentially small splitting of separatrices under fast quasiperiodic forcing*, Comm. Math. Phys. **189** (1997), 35–71.

[DGJS2] A. Delshams, V.G. Gelfreich, A. Jorba and T.M. Seara, *Lower and upper bounds for the splitting of separatrices of the pendulum under a fast quasiperiodic forcing*, Electron. Res. Announc. Amer. Math. Soc. **3** (1997), 1–10.

[DR1] A. Delshams and R. Ramírez-Ros, *Poincaré-Melnikov-Arnold method for analytic planar maps*, Nonlinearity **9** (1996), 1–26.

[DR2] A. Delshams and R. Ramírez-Ros, *Melnikov potential for exact symplectic maps*, Comm. Math. Phys. **190** (1997), 231–245

[DS1] A. Delshams and T.M. Seara, *An asymptotic expression for the splitting of separatrices of the rapidly forced pendulum*, Comm. Math. Phys. **150** (1992), 433–163.

[DS2] A. Delshams and T.M. Seara, *Splitting of separatrices in Hamiltonian systems with one and a half degrees of freedom*, Math. Phys. Electron. J. **3** (1997), 4–40.

[FS] B. Fiedler and J. Scheurle, *Discretization of homoclinic orbits, rapid forcing and "homoclinic" chaos*, Mem. Amer. Math. Soc. **119** (**570**) (1996), viii+79 pp.

[Fo1] E. Fontich, *Exponentially small upper bounds for the splitting of separatrices for high frequency periodic perturbations*, Nonlinear Anal. **20** (1993), 733–744.

[Fo2] E. Fontich, *Rapidly forced planar vector fields and splitting of separatrices*, J. Differential Equations **119** (1995), 310–335.

[Fo3] E. Fontich, *A flow box theorem for diffeomorphisms*, preprint, (2000).

[FS1] E. Fontich and C. Simó, *The splitting of separatrices for analytic diffeomorphisms*, Ergodic Theory Dynam. Systems **10** (1990), 295–318.

[FS2] E. Fontich and C. Simó, *Invariant manifolds for near identity differentiable maps and splitting of separatrices*, Ergodic Theory Dynam. Systems **10** (1990), 319–346.

[GGM] G. Gallavotti, G. Gentile and V. Mastropietro, *Separatrix splitting for systems with three times scales*, Comm. Math. Phys. **202** (**1**) (1999) 197–236.

[Ge1] V.G. Gelfreich, *Separatrices splitting for the rapidly forced pendulum*, Seminar on Dynamical Systems (St. Petersburg), Progress in Nonlinear Differential Equations and their Applications, vol. 12, Birkhäuser, 1991, pp. 47–67.

[Ge2] V.G. Gelfreich, *Melnikov method and exponentially small splitting of separatrices*, Phys. D **101** (1997), 227–248.

[Ge3] V.G. Gelfreich, *Reference systems for splitting of separatrices*, Nonlinearity **10** (1997), 175–193.

[Ge4] V.G. Gelfreich, *A proof of the exponentially small transversality of the separatrices for the standard map*, Comm. Math. Phys. **201** (**1**) (1999), 155–216.

[Ge5] V.G. Gelfreich, *Exponentially small splitting of separatrices for area-preserving maps*, Chaos, solitons and fractals **11** (2000) 241–243.

[GS] V.G. Gelfreich and D. Sauzin, *Borel summation and the splitting of separatrices of the Henón map*, Ann. Inst. Fourier (Grenoble) **51** (**2**) (2001), 513–567.

[HMS] P. Holmes, J. Marsden and J. Scheurle, *Exponentially small splitting of separatrices with applications to KAM theory and degenerate bifurcations*, Contemp. Math. vol. 81, 1988, pp. 213–244.

[La1] V.F. Lazutkin, *Splitting of separatrices for the Chirikov's standard map*, VINITI **6372/84** (1984). (English version in *www.maia.ub.es/mp_arc #98-421.*)

[La2] V.F. Lazutkin, *An analytic integral along the separatrix of the semistandard map: existence and an exponential estimate for the distance between the stable and unstable separatrices*, St. Petersburg Math. J. **4** (**4**) (1993), 721–748.

[LST] V.F. Lazutkin, I.G. Schachmanski and M.B. Tabanov, *Splitting of separatrices for standard and semistandard mappings*, Physica D **40** (1989), 235–348.

[Me] V.F. Melnikov, *On the stability of the center for time periodic perturbations*, Trans. Moscow Math. Soc. **12** (1963), 3–56.

[Ne] A.I. Neishtadt, *The separation of motions in systems with rapidly rotating phase*, J. Appl. Math. Mech. **48** (**2**) (1984), 133–139.

[Po] H. Poincaré, *Sur le problème des trois corps et les équations de la dynamique*, Acta Math. **13** (1890), 1–271.

[Sa] D. Sauzin, *A new method for measuring the splitting of invariant manifolds*, Annales Scientifiques de l' École Normale Supériore. Quatrième Série **34** (**2**) (2001), 159–221.

[SMH] J. Scheurle, J. Marsden and P. Holmes, *Exponentially small estimates for separatrix splitting*, Asymptotics beyond all orders (La Jolla, CA, 1991), NATO Adv. Sci. Inst. Ser. B Phys., vol. 284, Plenum, New York, 1991, pp. 187–195.

[Su] Y.B. Suris, *On the complex separatrices of some standard-like maps*, Nonlinearity **7** (1994), 1225–1236.

[Tr] D.V. Treschev, *Separatrix splitting for a pendulum with rapidly oscillating suspension point*, Russ. J. Math. Phys. **5** (**2**) (1997), 63–98.

Editorial Information

To be published in the *Memoirs*, a paper must be correct, new, nontrivial, and significant. Further, it must be well written and of interest to a substantial number of mathematicians. Piecemeal results, such as an inconclusive step toward an unproved major theorem or a minor variation on a known result, are in general not acceptable for publication. Papers appearing in *Memoirs* are generally longer than those appearing in *Transactions*, which shares the same editorial committee.

As of October 1, 2003, the backlog for this journal was approximately 5 volumes. This estimate is the result of dividing the number of manuscripts for this journal in the Providence office that have not yet gone to the printer on the above date by the average number of monographs per volume over the previous twelve months, reduced by the number of volumes published in four months (the time necessary for preparing a volume for the printer). (There are 6 volumes per year, each containing at least 4 numbers.)

A Consent to Publish and Copyright Agreement is required before a paper will be published in the *Memoirs*. After a paper is accepted for publication, the Providence office will send a Consent to Publish and Copyright Agreement to all authors of the paper. By submitting a paper to the *Memoirs*, authors certify that the results have not been submitted to nor are they under consideration for publication by another journal, conference proceedings, or similar publication.

Information for Authors

Memoirs are printed from camera copy fully prepared by the author. This means that the finished book will look exactly like the copy submitted.

The paper must contain a *descriptive title* and an *abstract* that summarizes the article in language suitable for workers in the general field (algebra, analysis, etc.). The *descriptive title* should be short, but informative; useless or vague phrases such as "some remarks about" or "concerning" should be avoided. The *abstract* should be at least one complete sentence, and at most 300 words. Included with the footnotes to the paper should be the 2000 *Mathematics Subject Classification* representing the primary and secondary subjects of the article. The classifications are accessible from `www.ams.org/msc/`. The list of classifications is also available in print starting with the 1999 annual index of *Mathematical Reviews*. The Mathematics Subject Classification footnote may be followed by a list of *key words and phrases* describing the subject matter of the article and taken from it. Journal abbreviations used in bibliographies are listed in the latest *Mathematical Reviews* annual index. The series abbreviations are also accessible from `www.ams.org/publications/`. To help in preparing and verifying references, the AMS offers MR Lookup, a Reference Tool for Linking, at `www.ams.org/mrlookup/`. When the manuscript is submitted, authors should supply the editor with electronic addresses if available. These will be printed after the postal address at the end of the article.

Electronically prepared manuscripts. The AMS encourages electronically prepared manuscripts, with a strong preference for AMS-LaTeX. To this end, the Society has prepared AMS-LaTeX author packages for each AMS publication. Author packages include instructions for preparing electronic manuscripts, the *AMS Author Handbook*, samples, and a style file that generates the particular design specifications of that publication series. Though AMS-LaTeX is the highly preferred format of TeX, author packages are also available in AMS-TeX.

Authors may retrieve an author package from e-MATH starting from `www.ams.org/tex/` or via FTP to `ftp.ams.org` (login as `anonymous`, enter username as password, and type `cd pub/author-info`). The *AMS Author Handbook* and the *Instruction Manual* are available in PDF format following the author packages link from `www.ams.org/tex/`. The author package can be obtained free of charge by sending email to `pub@ams.org` (Internet) or from the Publication Division, American Mathematical Society, 201 Charles St., Providence, RI 02904, USA. When requesting an author package, please specify AMS-LaTeX or AMS-TeX, Macintosh or IBM (3.5) format, and the publication in which your paper will appear. Please be sure to include your complete mailing address.

Sending electronic files. After acceptance, the source file(s) should be sent to the Providence office (this includes any TeX source file, any graphics files, and the DVI or PostScript file).

Before sending the source file, be sure you have proofread your paper carefully. The files you send must be the EXACT files used to generate the proof copy that was accepted for publication. For all publications, authors are required to send a printed copy of their paper, which exactly matches the copy approved for publication, along with any graphics that will appear in the paper.

TeX files may be submitted by email, FTP, or on diskette. The DVI file(s) and PostScript files should be submitted only by FTP or on diskette unless they are encoded properly to submit through email. (DVI files are binary and PostScript files tend to be very large.)

Electronically prepared manuscripts can be sent via email to `pub-submit@ams.org` (Internet). The subject line of the message should include the publication code to identify it as a Memoir. TeX source files, DVI files, and PostScript files can be transferred over the Internet by FTP to the Internet node `e-math.ams.org` (130.44.1.100).

Electronic graphics. Comprehensive instructions on preparing graphics are available at `www.ams.org/jourhtml/graphics.html`. A few of the major requirements are given here.

Submit files for graphics as EPS (Encapsulated PostScript) files. This includes graphics originated via a graphics application as well as scanned photographs or other computer-generated images. If this is not possible, TIFF files are acceptable as long as they can be opened in Adobe Photoshop or Illustrator. No matter what method was used to produce the graphic, it is necessary to provide a paper copy to the AMS.

Authors using graphics packages for the creation of electronic art should also avoid the use of any lines thinner than 0.5 points in width. Many graphics packages allow the user to specify a "hairline" for a very thin line. Hairlines often look acceptable when proofed on a typical laser printer. However, when produced on a high-resolution laser imagesetter, hairlines become nearly invisible and will be lost entirely in the final printing process.

Screens should be set to values between 15% and 85%. Screens which fall outside of this range are too light or too dark to print correctly. Variations of screens within a graphic should be no less than 10%.

Inquiries. Any inquiries concerning a paper that has been accepted for publication should be sent directly to the Electronic Prepress Department, American Mathematical Society, 201 Charles St., Providence, RI 02904, USA.

Editors

This journal is designed particularly for long research papers, normally at least 80 pages in length, and groups of cognate papers in pure and applied mathematics. Papers intended for publication in the *Memoirs* should be addressed to one of the following editors. In principle the Memoirs welcomes electronic submissions, and some of the editors, those whose names appear below with an asterisk (*), have indicated that they prefer them. However, editors reserve the right to request hard copies after papers have been submitted electronically. Authors are advised to make preliminary email inquiries to editors about whether they are likely to be able to handle submissions in a particular electronic form.

***Algebra** to ROBERT GURALNICK, Department of Mathematics, University of Southern California, Los Angeles, CA 90089-1113; email: `guralnic@math.usc.edu`

Algebraic geometry to DAN ABRAMOVICH, Department of Mathematics, Boston University, 111 Cummington St., Boston, MA 02215; email: `abramovic@bu.edu`

***Algebraic number theory** to V. KUMAR MURTY, Department of Mathematics, University of Toronto, 100 St. George Street, Toronto, ON M5S 1A1, Canada; email: `murty@math.toronto.edu`

Algebraic topology and cohomology of groups to STEWART PRIDDY, Department of Mathematics, Northwestern University, 2033 Sheridan Road, Evanston, IL 60208-2730; email: `priddy@math.nwu.edu`

Combinatorics and Lie theory to SERGEY FOMIN, Department of Mathematics, University of Michigan, Ann Arbor, Michigan 48109-1109; email: `fomin@umich.edu`

Complex analysis and complex geometry to DUONG H. PHONG, Department of Mathematics, Columbia University, 2990 Broadway, New York, NY 10027-0029; email: `phong@math.columbia.edu`

***Differential geometry and global analysis** to LISA C. JEFFREY, Department of Mathematics, University of Toronto, 100 St. George St., Toronto, ON Canada M5S 3G3; email: `jeffrey@math.toronto.edu`

Dynamical systems and ergodic theory to ROBERT F. WILLIAMS, Department of Mathematics, University of Texas, Austin, Texas 78712-1082; email: `bob@math.utexas.edu`

***Functional analysis and operator algebras** to MARIUS DADARLAT, Department of Mathematics, Purdue University, 150 N. University St., West Lafayette, IN 47907-2067; email: `mdd@math.purdue.edu`

***Geometric analysis** to TOBIAS COLDING, Courant Institute, New York University, 251 Mercer St., New York, NY 10012; email: `colding@cims.nyu.edu`

***Geometric analysis** to MLADEN BESTVINA, Department of Mathematics, University of Utah, 155 South 1400 East, JWB 233, Salt Lake City, Utah 84112-0090; email: `bestvina@math.utah.edu`

Harmonic analysis to ALEXANDER NAGEL, Department of Mathematics, University of Wisconsin, 480 Lincoln Drive, Madison, WI 53706-1313; email: `nagel@math.wisc.edu`

Harmonic analysis, representation theory, and Lie theory to ROBERT J. STANTON, Department of Mathematics, The Ohio State University, 231 West 18th Avenue, Columbus, OH 43210-1174; email: `stanton@math.ohio-state.edu`

***Logic** to STEFFEN LEMPP, Department of Mathematics, University of Wisconsin, 480 Lincoln Drive, Madison, Wisconsin 53706-1388; email: `lempp@math.wisc.edu`

Number theory to HAROLD G. DIAMOND, Department of Mathematics, University of Illinois, 1409 W. Green St., Urbana, IL 61801-2917; email: `diamond@math.uiuc.edu`

***Ordinary differential equations, and applied mathematics** to PETER W. BATES, Department of Mathematics, Michigan State University, East Lansing, MI 48824-1027; email: `peter@math.msu.edu`

***Partial differential equations** to PATRICIA E. BAUMAN, Department of Mathematics, Purdue University, West Lafayette, IN 47907-1395; email: `bauman@math.purdue.edu`

***Probability and statistics** to KRZYSZTOF BURDZY, Department of Mathematics, University of Washington, Box 354350, Seattle, Washington 98195-4350; email: `burdzy@math.washington.edu`

***Real analysis and partial differential equations** to DANIEL TATARU, Department of Mathematics, University of California, Berkeley, Berkeley, CA 94720; email: `tataru@ math.berkeley.edu`

All other communications to the editors should be addressed to the Managing Editor, WILLIAM BECKNER, Department of Mathematics, University of Texas, Austin, TX 78712-1082; email: `beckner@math.utexas.edu`.

Titles in This Series

795 **Adam Nyman,** Points on quantum projectivizations, 2004

794 **Kevin K. Ferland and L. Gaunce Lewis, Jr.,** The $RO(G)$-graded equivariant ordinary homology of G-cell complexes with even-dimensional cells for $G = \mathbb{Z}/p$, 2004

793 **Jindřich Zapletal,** Descriptive set theory and definable forcing, 2004

792 **Inmaculada Baldomá and Ernest Fontich,** Exponentially small splitting of invariant manifolds of parabolic points, 2004

791 **Eva A. Gallardo-Gutiérrez and Alfonso Montes-Rodríguez,** The role of the spectrum in the cyclic behavior of composition operators, 2004

790 **Thierry Lévy,** Yang-Mills measure on compact surfaces, 2003

789 **Helge Glöckner,** Positive definite functions on infinite-dimensional convex cones, 2003

788 **Robert Denk, Matthias Hieber, and Jan Prüss,** $\mathcal{R}$-boundedness, Fourier multipliers and problems of elliptic and parabolic type, 2003

787 **Michael Cwikel, Per G. Nilsson, and Gideon Schechtman,** Interpolation of weighted Banach lattices/A characterization of relatively decomposable Banach lattices, 2003

786 **Arnd Scheel,** Radially symmetric patterns of reaction-diffusion systems, 2003

785 **R. R. Bruner and J. P. C. Greenlees,** The connective K-theory of finite groups, 2003

784 **Desmond Sheiham,** Invariants of boundary link cobordism, 2003

783 **Ethan Akin, Mike Hurley, and Judy A. Kennedy,** Dynamics of topologically generic homeomorphisms, 2003

782 **Masaaki Furusawa and Joseph A. Shalika,** On central critical values of the degree four L-functions for GSp(4): The Fundamental Lemma, 2003

781 **Marcin Bownik,** Anisotropic Hardy spaces and wavelets, 2003

780 **S. Marmi and D. Sauzin,** Quasianalytic monogenic solutions of a cohomological equation, 2003

779 **Hansjörg Geiges,** h-principles and flexibility in geometry, 2003

778 **David B. Massey,** Numerical control over complex analytic singularities, 2003

777 **Robert Lauter,** Pseudodifferential analysis on conformally compact spaces, 2003

776 **U. Haagerup, H. P. Rosenthal, and F. A. Sukochev,** Banach embedding properties of non-commutative L^p-spaces, 2003

775 **P. Lochak, J.-P. Marco, and D. Sauzin,** On the splitting of invariant manifolds in multidimensional near-integrable Hamiltonian systems, 2003

774 **Kai A. Behrend,** Derived ℓ-adic categories for algebraic stacks, 2003

773 **Robert M. Guralnick, Peter Müller, and Jan Saxl,** The rational function analogue of a question of Schur and exceptionality of permutation representations, 2003

772 **Katrina Barron,** The moduli space of $N = 1$ superspheres with tubes and the sewing operation, 2003

771 **Shigenori Matsumoto,** Affine flows on 3-manifolds, 2003

770 **W. N. Everitt and L. Markus,** Elliptic partial differential operators and symplectic algebra, 2003

769 **Jie Wu,** Homotopy theory of the suspensions of the projective plane, 2003

768 **R. Höpfner and E. Löcherbach,** Limit theorems for null recurrent Markov processes, 2003

767 **Po Hu,** S-modules in the category of schemes, 2003

766 **Su Gao and Alexander S. Kechris,** On the classification of Polish metric spaces up to isometry, 2003

765 **Robert Bieri and Ross Geoghegan,** Connectivity properties of group actions on non-positively curved spaces, 2003

764 **J. Spandaw,** Noether-Lefschetz problems for degeneracy loci, 2003

763 **Yasuyuki Kachi and Eiichi Sato,** Segre's reflexivity and an inductive characterization os hyperquadrics, 2002

762 **Leiba Rodman, Ilya M. Spitkovsky, and Hugo Woerdeman,** Abstract band method via factorization, positive and band extensions of multivariable almost periodic matrix functions, and spectral estimation, 2002

761 **Oliver Druet and Emmanuel Hebey,** The AB program in geometric analysis : Sharp Sobolev inequalities and related problems, 2002

760 **Markus Banagl,** Extending intersection homology type invarients to non-Witt spaces, 2002

759 **Donald M. Davis,** From representation theory to homotopy groups, 2002

758 **Alan Forrest, John Hunton, and Johannes Kellendonk,** Topological invariants for projection method patterns, 2002

757 **Douglas Bowman,** q-difference operators, orthogonal polynomials, and symmetric expansions, 2002

756 **José Ignacio Cogolludo-Agustín,** Topological invariants of the complement to arrangements of rational plane curves, 2002

755 **M. A. Mandell and J. P. May,** Equivariant orthogonal spectra and S-modules, 2002

754 **Edward L. Green, Idun Reiten, and Øyvind Solberg,** Dualities on generalized Koszul algebras, 2002

753 **Daniel Panazzolo,** Desingularization of nilpotent singularities in families of planar vector fields, 2002

752 **Linus Kramer,** Homogeneous spaces, Tits buildings, and isoparametric hypersurfaces, 2002

751 **Bruce Allison, Georgia Benkart, and Yun Gao,** Lie algebras graded by the root systems BC_r, $r \geq 2$, 2002

750 **Masaki Izumi and Hideki Kosaki,** Kac algebras arising from composition of subfactors: General theory and classification, 2002

749 **Nanhua Xi,** The based ring of two-sided cells of affine Weyl groups of type $\widetilde{A}_{n-1}$, 2002

748 **Jürgen Ritter and Alfred Weiss,** The lifted root number conjecture and Iwasawa theory, 2002

747 **Armand Borel, Robert Friedman, and John W. Morgan,** Almost commuting elements in compact Lie groups, 2002

746 **Peter Niemann,** Some generalized Kac-Moody algebras with known root multiplicities, 2002

745 **Mikhail A. Lifshits and Werner Linde,** Approximation and entropy numbers of Volterra operators with application to Brownian motion, 2002

744 **Roger Chalkley,** Basic global relative invariants for homogeneous linear differential equations, 2002

743 **Heng Sun,** Spectral decomposition of a covering of $GL(r)$: the Borel case, 2002

742 **J. E. Gilbert, Y. S. Han, J. A. Hogan, J. D. Lakey, D. Weiland, and G. Weiss,** Smooth molecular functions and singular integral operators, 2002

741 **Francisco Santos,** Triangulations of oriented matroids, 2002

740 **Rick Durrett,** Mutual invadability implies coexistence in spatial models, 2002

739 **Georgios K. Alexopoulos,** Sub-Laplacians with drift on Lie groups of polynomial volume growth, 2002

For a complete list of titles in this series, visit the AMS Bookstore at **www.ams.org/bookstore/**.